ENVIRONMENTAL SCIENCE, ENGINEERING AND TECHNOLOGY

WILDFIRES AND WILDFIRE MANAGEMENT

ENVIRONMENTAL SCIENCE, ENGINEERING AND TECHNOLOGY

Nitrous Oxide Emissions Research Progress
Adam I. Sheldon and Edward P. Barnhart (Editors)
2009. ISBN: 978-1-60692-267-5

Fundamentals and Applications of Biosorption Isotherms, Kinetics and Thermodynamics
Yu Liu and Jianlong Wang (Editors)
2009. ISBN: 978-1-60741-169-7

Environmental Effects of Off-Highway Vehicles
Douglas S. Ouren, Christopher Haas, Cynthia P. Melcher, Susan C. Stewart, Phadrea D. Ponds, Natalie R. Sexton, Lucy Burris, Tammy Fancher and Zachary H. Bowen
2009. ISBN: 978-1-60692-936-0

Agricultural Runoff, Coastal Engineering and Flooding
Christopher A. Hudspeth and Timothy E. Reeve (Editors)
2009. ISBN: 978-1-60741-097-3
2009. ISBN: 978-1-60876-608-6 (E-book)

Conservation of Natural Resources
Nikolas J. Kudrow (Editor)
2009. ISBN: 978-1-60741-178-9
2009. ISBN: 978-1-60876-642-6 (E-book)

Directory of Conservation Funding Sources for Developing Countries: Conservation Biology, Education and Training, Fellowships and Scholarships
Alfred O. Owino and Joseph O. Oyugi
2009. ISBN: 978-1-60741-367-7

Forest Canopies: Forest Production, Ecosystem Health and Climate Conditions
Jason D. Creighton and Paul J. Roney (Editors)
2009. ISBN: 978-1-60741-457-5

Soil Fertility
Derek P. Lucero and Joseph E. Boggs (Editors)
2009. ISBN: 978-1-60741-466-7

Handbook of Environmental Policy
Johannes Meijer and Arjan der Berg (Editors)
2009. ISBN: 978-1-60741-635-7

Handbook of Environmental Research
Aurel Edelstein and Dagmar Bär (Editors)
2009. ISBN: 978-1-60741-492-6

Handbook on Environmental Quality
Evan K. Drury and Tylor S. Pridgen (Editors)
2009. ISBN: 978-1-60741-420-9
2009. ISBN: 978-1-61728-018-4 (E-book)

Environmental Cost Management
Randi Taylor Mancuso
2009. ISBN: 978-1-60741-815-3

The Amazon Gold Rush and Environmental Mercury Contamination
Daniel Marcos Bonotto and Ene Glória da Silveira
2009. ISBN: 978-1-60741-609-8

Biological and Environmental Applications of Gas Discharge Plasmas
Graciela Brelles-Mariño (Editor)
2009. ISBN: 978-1-60741-945-7

Buildings and the Environment
Jonas Nemecek and Patrik Schulz (Editors)
2009. ISBN: 978-1-60876-128-9

River Sediments
Greig Ramsey and Seoras McHugh (Editors)
2009. ISBN: 978-1-60741-437-7

Tree Growth: Influences, Layers and Types
Wesley P. Karam (Editor)
2009. ISBN: 978-1-60741-784-2

Sorbents: Properties, Materials and Applications
Thomas P. Willis (Editor)
2009. ISBN: 978-1-60741-851-1
2009. ISBN: 978-1-61668-308-5 (E-book)

Syngas: Production Methods, Post Treatment and Economics
Adorjan Kurucz and Izsak Bencik (Editors)
2009. ISBN: 978-1-60741-841-2
2009. ISBN: 978-1-61668-214-9 (E-book)

Process Engineering in Plant-Based Products
Hongzhang Chen
2009. ISBN: 978-1-60741-962-4

Potential of Activated Sludge Utilization
Xiaoyi Yang
2009. ISBN: 978-1-60876-019-0

Recent Progress on Earthquake Geology
Pierpaolo Guarnieri (Editor)
2009. ISBN: 978-1-60876-147-0

Estimating Future Recreational Demand
Peter T. Yao (Editor)
2009. ISBN: 978-1-60692-472-3

Bioengineering for Pollution Prevention
Dianne Ahmann and John R. Dorgan
2009. ISBN: 978-1-60692-900-1
2009. ISBN: 978-1-60876-574-4 (E-book)

Floodplains: Physical Geography, Ecology and Societal Interactions
Marc A. Álvarez (Editor)
2010. ISBN: 978-1-61728-277-5
2010. ISBN: 978-1-61728-608-7 (E-book)

Aquifers: Formation, Transport and Pollution
Rachel H. Laughton (Editor)
2010. ISBN: 978-1-61668-051-0
2010. ISBN: 978-1-61668-444-0 (E-book)

Temporarily Open/Closed Estuaries in South Africa
R. Perissinotto, D.D. Stretch, A.K. Whitfield, J.B. Adams,
A.T. Forbes and N.T. Demetriades
2010. ISBN: 978-1-61668-412-9
2010. ISBN: 978-1-61668-825-7 (E-book)

Carbon Capture and Storage including Coal-Fired Power Plants
Todd P. Carington (Editor)
2010. ISBN: 978-1-60741-196-3

Handbook on Agroforestry: Management Practices and Environmental Impact
Lawrence R. Kellimore (Editor)
2010. ISBN: 978-1-60876-359-7

Biodiversity Hotspots
Vittore Rescigno and Savario Maletta (Editors)
2010. ISBN: 978-1-60876-458-7

Mine Drainage and Related Problems
Brock C. Robinson (Editor)
2010. ISBN: 978-1-60741-285-4
2010. ISBN: 978-1-61668-643-7 (E-book)

Pipelines for Carbon Sequestration: Background and Issues
Elvira S. Hoffmann (Editor)
2010. ISBN: 978-1-60741-383-7

The Role of Forests in Carbon Capture and Climate Change
Roland Carnell (Editor)
2010. ISBN: 978-1-60741-447-6

Clean Fuels in the Marine Sector
Environmental Protection Agency
2010. ISBN: 978-1-60741-275-5
2010. ISBN: 978-1-61668-431-0 (E-book)

Freshwater Ecosystems and Aquaculture Research
Felice De Carlo and Alessio Bassano (Editors)
2010. ISBN: 978-1-60741-707-1

Species Diversity and Extinction
Geraldine H. Tepper (Editor)
2010. ISBN: 978-1-61668-343-6
2010. ISBN: 978-1-61668-406-8 (E-book)

Harmful Algal Blooms – Impact and Response
Vladimir Buteyko (Editor)
2010. ISBN: 978-1-60741-665-4

Estuaries: Types, Movement Patterns and Climatical Impacts
Julian R. Crane and Ashton E. Solomon (Editors)
2010. ISBN: 978-1-60876-859-2

Built Environment: Design, Management and Applications
Paul S. Geller (Editor)
2010. ISBN: 978-1-60876-915-5

Wildfires and Wildfire Management
Kian V. Medina (Editor)
2010. ISBN: 978-1-60876-009-1

HYDRO GIS: Theory and Lessons from the Vietnamese Delta
Shigeko Haruyama and Le Thie Viet Hoa
2010. ISBN: 978-1-60876-156-2

Grassland Biodiversity: Habitat Types, Ecological Processes and Environmental Impacts
Johan Runas and Theodor Dahlgren (Editors)
2010. ISBN: 978-1-60876-542-3

Check Dams, Morphological Adjustments and Erosion Control in Torrential Streams
Carmelo Consesa Garcia and Mario Aristide Lenzi (Editors)
2010. ISBN: 978-1-60876-146-3

Fluid Waste Disposal
Kay W. Canton (Editor)
2010. ISBN: 978-1-60741-915-0

Psychological Approaches to Sustainability: Current Trends in Theory, Research and Applications
Victor Corral-Verdugo, Cirilo H. Garcia-Cadena and Martha Frias-Armenta (Editors)
2010. ISBN: 978-1-60876-356-6

Environmental Modeling with GPS
Lubos Matejicek (Editor)
2010. ISBN: 978-1-60876-363-4

Pollen: Structure, Types and Effects
Benjamin J. Kaiser (Editor)
2010. ISBN: 978-1-61668-669-7
2010. ISBN: 978-1-61728-048-1 (E-book)

Mechanisms of Cadmium Toxicity to Various Trophic Saltwater Organisms
Zaosheng Wang, Changzhou Yan, Hainan Kong and Deyi Wu
2010. ISBN: 978-1-60876-646-8

Wood: Types, Properties, and Uses
Lorenzo F. Botannini (Editor)
2010. ISBN: 978-1-61668-837-0
2010. ISBN: 978-1-61728-046-7 (E-book)

Eco-City and Green Community: The Evolution of Planning Theory and Practice
Zhenghong Tang (Editor)
2010. ISBN: 978-1-60876-811-0

Anthropology of Mining in Papua New Guinea Greenfields
Benedict Young Imbun
2010. ISBN: 978-1-61668-485-3

Paleoecological Significance of Diatoms in Argentinean Estuaries
Gabriela S. Hassan
2010. ISBN: 978-1-60876-953-7

Geomatic Solutions for Coastal Environments
M. Maanan and M. Robin (Editors)
2010. ISBN: 978-1-61668-140-1

Marine Research and Conservation in the Coral Triangle: The Wakatobi National Park
Julian Clifton, Richard K.F. Unsworth and David J. Smith (Editors)
2010. ISBN: 978-1-61668-473-0

Natural Resources in Ghana: Management, Policy and Economics
David M. Nanang and Thompson K. Nunifu (Editors)
2010. ISBN: 978-1-61668-020-6

Modelling Flows in Environmental and Civil Engineering
F. Kerger, B.J. Dewals, S. Erpicum, P. Archambeau and M. Pirotton
2010. ISBN: 978-1-61668-662-8
2010. ISBN: 978-1-61668-490-7 (E-book)

Fundamentals of General Ecology, Life Safety and Environment Protection
Mark D. Goldfein, Alexei V. Ivanov and Nikolaj V. Kozhevnikov
2010. ISBN: 978-1-61668-176-0
2010. ISBN: 978-1-61668-397-9 (E-book)

Zinc, Copper, or Magnesium Supplementation Against Cadmium Toxicity
Vesna Matović, Zorica Plamenac Bulat, Danijela Đukić-Ćosić and Danilo Soldatović
2010. ISBN: 978-1-61668-332-0
2010. ISBN: 978-1-61668-721-2 (E-book)

Sources and Reduction of Greenhouse Gas Emissions
Steffen D. Saldana (Editor)
2010. ISBN: 978-1-61668-856-1
2010. ISBN: 978-1-61728-091-7 (E-book)

International Trade and Environmental Justice: Toward a Global Political Ecology
Alf Hornborg and Andrew K. Jorgenson (Editors)
2010. ISBN: 978-1-60876-426-6

Amazon Basin: Plant Life, Wildlife and Environment
Nicolás Rojas and Rafael Prieto (Editors)
2010. ISBN: 978-1-60741-463-6

Biogeography
Mihails Gailis and Stefans Kalniòð (Editors)
2010. ISBN: 978-1-60741-494-0

Behavioral and Chemical Ecology
Wen Zhang and Hong Liu (Editors)
2010. ISBN: 978-1-60741-099-7

Global Environmental Policies: Impact, Management and Effects
Riccardo Cancilla and Monte Gargano (Editors)
2010. ISBN: 978-1-60876-204-0

Tundras: Vegetation, Wildlife and Climate Trends
Beltran Gutierrez and Cristos Pena (Editors)
2010. ISBN: 978-1-60876-588-1

Advanced Biologically Active Polyfunctional Compounds and Composites: Health, Cultural Heritage and Environmental Protection
Nodar Lekishvili, Gennady Zaikov and Bob Howell (Editors)
2010. ISBN: 978-1-60876-114-2

How Globalization is Changing the U.S. Forest Sector
Peter Ince, Albert Schuler, Henry Spelter and William Luppold (Editors)
2010. ISBN: 978-1-60876-132-6

Environmental Modeling with GPS
Lubos Matejicek (Editor)
2010. ISBN: 978-1-60876-363-4

Watersheds: Management, Restoration and Environmental Impact
Jeremy C. Vaughn (Editor)
2010. ISBN: 978-1-61668-667-3
2010. ISBN: 978-1-61728-243-0 (E-book)

A True Tale of Science and Discovery
Lawrence A. Curtis
2010. ISBN: 978-1-60876-595-9

Advances in Environmental Modeling and Measurements
Dragutin T. Mihailovic and Branislava Lalic (Editors)
2010. ISBN: 978-1-60876-599-7

Alaskan Native Villages Threatened by Erosion
Russell M. Trevino (Editor)
2010. ISBN: 978-1-60876-890-5

Aviation and Climate Change
George T. Blumenthal (Editor)
2010. ISBN: 978-1-60876-757-1

Protecting the Great Lakes from Invasive and Nonindigenous Species
Clara E. Wouters (Editor)
2010. ISBN: 978-1-61728-103-7
2010. ISBN: 978-1-61728-330-7 (E-book)

Wildfires, Fuels and Invasive Plants
Louise E. Willems (Editor)
2010. ISBN: 978-1-61728-164-8
2010. ISBN: 978-1-61728-322-2 (E-book)

Spatial Assemblages of Tropical Intertidal Rocky Shore Communities in Ghana, West Africa
Emmanuel Lamptey, Ayaa Kojo Armah and Lloyd Cyril Allotey
2010. ISBN: 978-1-61668-767-0
2010. ISBN: 978-1-61728-448-9 (E-book)

ENVIRONMENTAL SCIENCE, ENGINEERING AND TECHNOLOGY

WILDFIRES AND WILDFIRE MANAGEMENT

KIAN V. MEDINA
EDITOR

Nova Science Publishers, Inc.
New York

Library of Congress Cataloging-in-Publication Data

Wildfires and wildfire management / editor: Kian V. Medina.
p. cm.
Includes index.
ISBN 978-1-60876-009-1 (hardcover)
1. Wildfires. 2. Fire management. 3. Wildfires--California. 4. Fire management--California. I. Medina, Kian V.
SD421.W5454 2010
363.37'9--dc22

2009046497

Published by Nova Science Publishers, Inc. ✦ New York

CONTENTS

PREFACE

Congress continues to face questions about forestry practices, funding levels, and the federal role in wildland fire protection. Recent fire seasons have been, by most standards, among the worst in the past half century. In addition, many factors contribute to the threat of wildlife damages. Two major factors are the decline in forest and rangeland health and the expansion of residential areas into wildlands. It should also be recognized that, as long as biomass, drought and high winds exist, catastrophic wildfires will occur. This book presents results of completed questionnaires and focus group comments organized around reactions to and beliefs about wildfires and wildfire management. This book also focuses on options for protecting structures and for protecting wildlands and natural resources from wildfires. The book begins with a brief overview of the nature of wildfires, followed by a discussion of protecting structures. Wildlife damages to wildlands and natural resources are also discussed, as well as fuel treatment options and their benefits and limitations, and public involvement in federal decisions. This book consists of public documents which have been located, gathered, combined, reformatted, and enhanced with a subject index, selectively edited and bound to provide easy access.

Chapter 1 - Generally speaking, losses from wildfires have been a manageable risk in the private insurance market. Insurance coverage has been widely available both in the standard insurance market and in residual or "involuntary" markets established through state legislation in the early 1970s to assure markets for risks not always available or affordable in the standard market. In California, applicants for fire coverage under the Fair Access to Insurance Requirement (FAIR) plan must live in areas of the state specifically designated by the insurance commissioner. Assistance for uninsured losses is being met through standing authorities; the need for additional federal legislation is not yet known. Federal costs to cover uninsured losses associated with the wildfires in California may require supplemental appropriations. Some may also argue that pending legislation (H.R. 3355/S. 2310, the Homeowners' Defense Act of 2007) would provide a federal backstop for state-sponsored insurance programs to help homeowners prepare for and recover from the damages caused by natural catastrophes such as the wildfires.

Chapter 2 - This chapter presents results from a study of San Bernardino National Forest community residents' experiences with and perceptions of fire, fire management, and the Forest Service. Using self-administered surveys and focus group discussions, we found that participants had personal experiences with fire, were concerned about fire, and felt knowledgeable about effective fire management. Consideration of future consequences, a measure of time orientation, was not found to be related to beliefs about and reactions to wildfire. Trust in the Forest Service was related to a number of fire-associated attitudes. Findings help shed light on the experiences of residents living in fire-prone communities and highlight the importance of trust in understanding public perceptions about fire management.

Chapter 3 - Congress continues to face questions about forestry practices, funding levels, and the federal role in wildland fire protection. Recent fire seasons have been, by most standards, among the worst in the past half century. National attention began to focus on wildfires when a prescribed burn in May 2000 escaped control and burned 239 homes in Los Alamos, NM. President Clinton responded by requesting a doubling of wildfire management funds, and Congress enacted much of this proposal in the FY2001 Interior appropriations act (P.L. 106-291). President Bush responded to the severe 2002 fires by proposing a Healthy Forests Initiative to reduce fuel loads by expediting review processes.

Chapter 4 - Wildfires are getting more severe, with more acres and houses burned and more people at risk. This results from excess biomass in the forests, due to past logging and grazing and a century of fire suppression, combined with an expanding wildlandurban interface — more people and houses in and near the forests — along with climate change, exacerbating drought and insect and disease problems. Some assert that current efforts to reduce biomass (fuel treatments, such as thinning) and to protect houses are inadequate, and that public objections to activities on federal lands raise costs and delay action. Others counter that proposals for federal lands allow timber harvesting, with substantial environmental damage and little fire protection. Congress is addressing these issues through various legislative proposals and through funding for protection programs.

Chapter 5 - Congress is giving increased attention and funding to wildfire threats. Much of the concern focuses on protecting homes and other structures in and near forests, an area known as the *wildland-urban interface*. However, not all agree on what can and should be done during wildfires, in their aftermath, and especially beforehand to protect the interface. This chapter describes the growth of the wildland-urban interface, wildfire suppression efforts, post-fire responses, and especially the programs and options for protecting the interface before the next wildfire strikes.

Wildfires have made national headlines in recent years, with major fires in the West and South killing firefighters, burning homes, and threatening communities. Federal funding for fire protection has more than doubled in the past decade, and administration and congressional leaders have urged additional wildfire protection. (See CRS Report RL33990, *Wildfire Funding*, by Ross W. Gorte.) Attention has focused on protecting people, homes, and communities in the *wildland-urban interface* (WUI), but opinions vary over how to protect the interface.

Chapter 6 - Wildland fires have become increasingly damaging and costly. To deal with fire's threats, the five federal wildland fire agencies—the Forest Service in the Department of Agriculture and four agencies in the Department of the Interior (Interior)—rely on thousands of firefighters, fire engines, and other assets. To ensure acquisition of the best mix of these assets, the agencies in 2002 began developing a new interagency budget tool known as fire program analysis (FPA). FPA underwent major changes in 2006, raising questions about its ability to meet its original objectives. GAO was asked to examine (1) FPA's development to date, including the 2006 changes, and (2) the extent to which FPA will meet its objectives. To do so, GAO reviewed agency policies and FPA documentation and interviewed agency officials.

In: Wildfires and Wildfire Management
Editor: Kian V. Medina

ISBN: 978-1-60876-009-1

Chapter 1

CALIFORNIA WILDFIRES: THE ROLE OF DISASTER INSURANCE

Rawle O. King

ABSTRACT

Generally speaking, losses from wildfires have been a manageable risk in the private insurance market. Insurance coverage has been widely available both in the standard insurance market and in residual or "involuntary" markets established through state legislation in the early 1970s to assure markets for risks not always available or affordable in the standard market. In California, applicants for fire coverage under the Fair Access to Insurance Requirement (FAIR) plan must live in areas of the state specifically designated by the insurance commissioner. Assistance for uninsured losses is being met through standing authorities; the need for additional federal legislation is not yet known. Federal costs to cover uninsured losses associated with the wildfires in California may require supplemental appropriations. Some may also argue that pending legislation (H.R. 3355/S. 2310, the Homeowners' Defense Act of 2007) would provide a federal backstop for state-sponsored insurance programs to help homeowners prepare for and recover from the damages caused by natural catastrophes such as the wildfires.

SUMMARY

The tragic consequences of the wildfires that struck southern California in late October 2007, have given renewed attention to the partnership between private providers of disaster insurance and the federal government. In broad terms, the disruption to economic systems caused by natural disasters, such as wildfires, windstorms, earthquakes, and floods, have been handled by the insurance and reinsurance industries and by the federal government (taxpayers). Consequently, large government outlays for disaster assistance and higher premiums for disaster insurance and reinsurance have followed the devastation caused by natural and man-made disasters. While it is too early to determine the full impact of the 2007 California wildfires on state and national property insurance markets, early estimates of $1 billion in insured property losses suggest this event will not exceed the most destructive fire in the state's history — the 1991 Oakland fires that cost $2.5 billion in 2006 dollars. The scope of the losses is well within the demonstrated capacity of the private insurance and reinsurance markets.

Generally speaking, losses from wildfires have been a manageable risk in the private insurance market. Insurance coverage has been widely available both in the standard insurance market and in residual or "involuntary" markets established through state legislation in the early 1970s to assure markets for risks not always available or affordable in the standard market. In California, applicants for fire coverage under the Fair Access to Insurance Requirement (FAIR) plan must live in areas of the state specifically designated by the insurance commissioner.

Assistance for uninsured losses is being met through standing authorities; the need for additional federal legislation is not yet known. Federal costs to cover uninsured losses associated with the wildfires in California may require supplemental appropriations. Some may also argue that pending legislation (H.R. 3355/S. 2310, the Homeowners' Defense Act of 2007) would provide a federal backstop for state-sponsored insurance programs to help homeowners prepare for and recover from the damages caused by natural catastrophes such as the wildfires.

WILDFIRE DATA OVERVIEW

On October 24, 2007, President Bush issued a federal emergency disaster declaration in response to property damage from wind-driven southern California wildfires that destroyed approximately 2,200 homes on 450,000 acres of land in seven counties stretching from Los Angeles to San Diego.[1] As of October 24, 2007, estimates of property damage provided by the Insurance Information Network of California suggests that insurance claims "will likely top $1 billion."[2]

While California has a history of significant outbreaks of wildfire, such disasters are not commonplace, nor do they consistently exceed the billion dollar threshold in insured losses that the current wildfires are expected to reach. **Table 1** shows that the current blazes remain short of the $2.5 billion in insured losses (2006 dollars) from the state's worst series of wildfires in over thirty years, the 1991 Oakland/Alameda fire that destroyed 2,000 homes.

Based on available information, it appears that the current wildfires may not result in the over $2 billion (2006 dollars) in losses that resulted from the wildfires that swept through southern California in 2003, causing $1.2 billion in insured losses in San Diego County and $1.1 billion in San Bernardino County. According to the data compiled by the insurance industry, since 2003 California had been relatively calm in terms of catastrophic property losses until the current outbreak.

Table 1. Top Ten Catastrophic Wild land Fires in California, 1970-2007 (ranked by cost in millions of 2006 dollars)

Rank	Date	Location	Nominal Dollars	Real 2006 Dollars
1	Oct. 20-21, 1991	Oakland, Alameda Counties, CA	$1,700	$2,516.3
2	Oct. 25- Nov. 4, 2003	San Diego County, CA	1,060	1,161.4
3	Oct. 25-Nov. 3, 2003	San Bernardino County, CA	975	1,068.3
4	Nov. 2-3, 1993	Los Angeles County, CA	375	523.2
5	Oct. 27-28, 1993	Orange County, CA	30	488.3
6	June 27-July 2, 1990	Santa Barbara County, CA	265	408.8
7	July 2007	Lake Tahoe	150	150.0
8	Sep. 22-30, 1970	Oakland-Berkeley Hills, CA	24.8	128.9
9	Nov. 24-30, 1980	Los Angeles, San Bernardino, Orange, Riverside, San Diego Counties, CA	43	105.2
10	July 26-27, 1977	Santa Barbara, Montecito, CA	20	66.5

Source: Insurance Services Office's Property Claims Services, Insurance Information Institute.

INSURANCE AND DISASTER RECOVERY

In brief, when a disaster such as the 2007 California wildfire occurs, productive components in the stricken region or state are destroyed. Resources from other parts of the country are redirected to compensate the victims and to rebuild what was lost or repair what was damaged. The impact of the disaster on the affected region, and its economic recovery, depend in large part on the portion of the losses covered by insurance, or reimbursed through public post-disaster assistance.[3] Through insurance, the risk of financial loss is transferred to an insurance company or other insuring organization.

The problem of under-insurance has been an issue before Congress. A home is considered under-insured when the homeowner purchases less insurance protection than is needed to rebuild after a disaster. Several reasons could explain why a homeowner is under-insured: he or she may not update the policy coverage limits; receive incorrect advice from an insurance agent or insurer about how much insurance is needed; or possess limited financial resources to buy the appropriate amount of coverage. After Hurricanes Isabel in 2003 and Katrina in 2005, for example, homeowners from affected states expressed concerns about claims payments that were substantially lower than the actual replacement value of their homes. Their reasons for under-insurance were all of the above. Likewise, homeowners affected by the California wildfires reportedly have also found the cost of rebuilding exceeds the available insurance claims payments. When property owners have insurance coverage, the uninsured portion of the losses must be absorbed by the property owner, unless federal assistance provides supplementary help.

During the disaster recovery period, the affected local community engages in redevelopment and cleanup efforts (assuming a willingness on the part of investors to redevelop the area after a disaster) that tend to increase local employment and other economic activities. As a general rule, insurance payments and disaster assistance provide a flow of funds into the area.[4] Realizing the potential for profits, investors will likely be attracted to the building boom in the devastated area.

Insurers are able to assume their policyholders' risk because they insure many individuals and can rely upon the law of large numbers to ensure profitability. Insurers know that when a large number of individual risks are combined, the total amount of loss can be predicted with reasonable accuracy. The insurance industry's expertise, therefore, is to forecast the total amount of loss payments for an entire group of risks and charge an amount to cover losses plus the cost of operating the insurance business and a margin for profit. The cost of operating the business is then divided among all the policyholders.

Damages caused by fire and smoke are generally covered under the standard homeowners, renters, and business insurance policies and under the comprehensive portion of an auto insurance policy. While Congress established the National Flood Insurance Program in 1968 due to the absence and high cost of private insurance, no similar program has been established for fire losses, as the private sector generally provides sufficient coverage. Congress and state legislatures have acted in the past, however, to ensure that homeowners in areas deemed higher risk are able to purchase insurance. For example, in addition to standard insurance, homeowners in wildfire-prone areas of California may also obtain insurance protection from California's Fair Access to Insurance Requirement (FAIR) plan.

FAIR ACCESS TO INSURANCE RATE (FAIR) PLANS

During the urban riots and civil disorders of the 1960s, many of America's cities suffered property losses that caused many private insurers to become reluctant to underwrite property insurance in communities with a high potential for loss. As property insurers withdrew from inner city neighborhoods, citing huge losses, Congress passed, as part of the Housing and Urban Development Act of 1968, three major property insurance programs to alleviate the availability and affordability problem. Fair Access to Insurance Requirement (FAIR) plans provided insurance against fire, riot, and looting to homeowners and businesses who could not obtain such insurance in the voluntary market. FAIR plans are essentially state-mandated and supervised insurance pools of all property insurance companies operating within a state. Although the FAIR plans act as a single insurer, participating insurers actually share the premiums as well as the profits or losses and expenses incurred, based on their share of the voluntary property market in the state. Property owners residing in eligible urban communities (or in designated hazardous brush areas) who are able to meet reasonable underwriting standards, such as minimum fire and health protection standards, may apply to their state's FAIR plan for coverage. FAIR plans offer rates that are set at break-even levels.

As of July 2007, 32 states and the District of Columbia had FAIR plans. In most states, subsidies lower rates below the amount that would be charged in the voluntary market for the same level of risk. While most FAIR plans have lost money in high loss years, they have also been profitable in other years. Losses under the FAIR plan are covered by assessment imposed on member insurers according to their share of the voluntary property market in the state. Insurers are then able to pass the losses onto policyholders in the form of higher rates and in some states to policyholders in the voluntary market as well. Several states allow insurers to recoup losses through rate surcharges. These charges, however, are itemized on a policyholder's premium bill.

In all states except California, residents in any part of the state can apply for insurance through the FAIR Plan as long as they meet Plan criteria. In California, applicants for fire coverage must live in areas specifically designated by the insurance commissioner. These include not only urban communities and some entire counties but also certain areas that are prone to brush fires. According to the Property Insurance Plans Service Office (PIPSO), an organization that compiles data and information on FAIR plans nationwide, in 2006, the California FAIR plan generated $82 million in direct written premiums from 193,615 habitational policies and 12,509 commercial policies with exposure of $51 billion.[5]

CONCLUSION

Everyone who needs insurance coverage, particularly those in high-risk disaster- prone areas, might not be able to buy insurance because of its cost or the lack of availability. As a result, a number of residual market mechanisms have been established to meet the needs of these buyers. Generally speaking, insured losses from wildfires, however, have been a manageable risk, unlike, say, hurricane risks in most Gulf and Atlantic coast states — particularly following Hurricanes Katrina, Rita, and Wilma in 2005. Insurance coverage for wildfires is widely available both in the standard insurance market and residual or "involuntary" markets established by acts of state legislatures in the early 1970s to assure a market for risks that found difficulty obtaining property coverage the standard market. In California, applicants for fire coverage provided under the FAIR plan must live in areas of the state specifically designated by the insurance commissioners.

Legislative Options

Should the current appropriations for federal disaster assistance prove insufficient to meet the needs of California, Congress may face a request to appropriate additional funds. In addition, Members of the 110th Congress might elect to address the issue of catastrophe insurance through pending legislation such as H.R. 3355/S. 2310, the Homeowners' Defense Act of 2007. These identical bills would establish a nonprofit National Catastrophe Risk Consortium authorized to inventory catastrophe risk obligations held by participating state reinsurance funds, risk pools, or primary insurance corporations; to issue securities and other financial instruments linked to catastrophe risk in the capital markets; or to enter into reinsurance contracts with private parties, among other purposes.

End Notes

[1] "Major Disaster Declarations, California Wildfires," at [http://www.fema.gov/news/ event.fema?id=9045], visited October 25, 2007.

[2] Insurance Information Network of California, "October 24: Insurance Claims Filings and Estimated Insured Losses," at [http://www.iinc.org/articles/22 1/1/October-24 — InsuranceClaims-Filings-and-Estimated-Insured-Losses/Page 1 .html], visited October 25, 2007.

[3] Information on federal disaster assistance is available in CRS Report RL3 3053, *Federal Stafford Act Disaster Assistance: Presidential Declarations, Eligible Activities, and Funding*, by Keith Bea. Information on federal funding through supplemental appropriations is presented in CRS Report RL33226, *Emergency Supplemental Appropriations Legislation for Disaster Assistance: Summary Data, FY1989 to FY2007*, by Justin Murray and Keith Bea.

[4] Certain catastrophic disasters have led some insurers to reconsider past practices to address unanticipated losses. For background see CRS Report RL32825, *Hurricanes and Disaster Risk Financing Through Insurance: Challenges and Policy Options*, by Rawle O. King.

[5] Information on PIPSO is available at [http://www.pipso.com/], visited October 24, 2007. Users of the website must be authorized to have access.

In: Wildfires and Wildfire Management
Editor: Kian V. Medina

ISBN: 978-1-60876-009-1

Chapter 2

THE EXPERIENCE OF COMMUNITY RESIDENTS IN A FIRE-PRONE ECOSYSTEM: A CASE STUDY ON THE SAN BERNARDINO NATIONAL FOREST

George T. Cvetkovich and Patricia L. Winter

ABSTRACT

Cvetkovich, George T.; Winter, Patricia L. 2008. The experience of community residents in a fire-prone ecosystem: a case study on the San Bernardino National Forest. Res. Pap. PSW-RP-257. Albany, CA: U.S. Department of Agriculture, Forest Service, Pacific Southwest Research Station. 42 p.

This chapter presents results from a study of San Bernardino National Forest community residents' experiences with and perceptions of fire, fire management, and the Forest Service. Using self-administered surveys and focus group discussions, we found that participants had personal experiences with fire, were concerned about fire, and felt knowledgeable about effective fire management. Consideration of future consequences, a measure of time orientation, was not found to be related to beliefs about and reactions to wildfire. Trust in the Forest Service was related to a number of fire-associated attitudes. Findings help shed light on the experiences of residents living in fire-prone communities and highlight the importance of trust in understanding public perceptions about fire management.

Keywords: Fire-prone communities, San Bernardino National Forest, fire management, trust, salient values similarity.

SUMMARY

Residents of fire-prone communities proximate to and surrounded by the San Bernardino National Forest participated in this study. As a group, the participants are characterized as having personally experienced wildland-fire-related events, being highly concerned about fires and fire risks, and self-assessed as knowledgeable about what effective fire management should be. This chapter presents results of completed questionnaires and focus group comments organized around reactions to and beliefs about wildfires and wildfire management. These are (1) personal stress-related consequences of directly experiencing wildfires and living in communities threatened by wildfires, concern about the risk of wildfires, and assessments of level of knowledge about wildfire management; (2) perceived level of responsibility for wildfire prevention, participation in fire management activities, and perceived barriers to effective fire management; and (3) views about preferred ways of receiving communication and education about wildfires and management. The report also presents analyses of the relationship between these reactions and beliefs and two measures of individual differences. Consideration of future consequences, a measure of individual differences in future time perspective found to be associated with differences in environmental attitudes and behaviors in previous research, was not strongly correlated to wildfire- related reactions and beliefs. High trust of the Forest Service was related to having fewer direct experiences with fire and related stress reactions, giving the Forest Service a high grade for efforts to prevent fires in the past year, and agreeing that the past record of fire management was a good reason to rely on the Forest Service. Participants trusting the Forest Service also agreed that the Forest Service shared their values for wildland fire management, that the Forest Service's management actions had been consistent with shared values, and that any value/action inconsistencies were justified.

INTRODUCTION

Conditions in the national forests resulting from drought, bark beetle infestation, abundant fuel supplies owing to fire suppression, high tree densities, and arson (Molloy 2004) have resulted in a high threat of wildland fire. In 2003, one expert summarized the destruction from wildfires since 1990 as follows:

> [W]e have lost 50 million acres of forest to wildfire and suffered the destruction of over 4,800 homes. The fires of 2000 burned 8.4 million acres and destroyed 861 structures. The 2002 fire season resulted in a loss of 6.9 million acres and 2,381 structures destroyed, including 835 homes. These staggering losses from wildfire also resulted in taxpayers paying $2.9 billion in firefighting costs. This does not include vast sums spent to rehabilitate damaged forests and replace homes [Bonnicksen 2003].

Since then, the San Bernardino National Forest, the focus of this study, has experienced major fires such as the Old Fire in 2003 and the Esperanza Fire in 2006. These and other fires have added to the toll through the burning of hundreds of thousands of additional acres of forest, the destruction of hundreds of homes and other property, the loss of human lives, and a high cost for firefighting (U.S. Department of Agriculture Inspector General 2006). The communities included in this study are adjacent to the national forest and other federal lands and have been listed by the California Department of Forestry and Fire Protection as Hazard Level Code "3," indicating the highest fire threat level (Inland Empire Fire Safe Alliance 2006).

This study examines three broadly grouped sets of reactions to and beliefs about wildfires and wildfire management: (1) personal stress-related consequences of directly experiencing wildland fires and living in communities threatened by wildfires, concern about the risk of wildland fires, and assessments of level of knowledge about wildfire management; (2) perceived level of responsibility for wildfire prevention, participation in fire management activities, and perceived barriers to effective fire management; and (3) views about preferred ways of receiving communication and education about wildfires and management. This information is presented in two parts: (1) results relating to reactions and beliefs about wildfires and (2) relationship of the reactions and beliefs to two factors–consideration of future consequences and trust of the USDA Forest Service (and associated measures).

Consideration of Future Consequences

Consideration of future consequences is a form of future time perspective that motivates an individual's efforts to reach desirable outcomes by focusing either on distant or immediate consequences of potential behaviors (Strathman et al. 1994). A reliable and valid measure of consideration of future consequences has been developed and was used in this study (Joireman 1999, Joireman et al. 2004, Petrocelli 2003). Individual differences in level of consideration of future consequences have been found to be related to various health and environment- related attitudes and behaviors. These include attitudes concerning private automobiles versus public transportation (Collins and Chambers 2005; Joireman et al. 2001, 2004), recycling and waste reduction (Ebreo and Vining 2001), sensitivity to health communications (Orbell et at. 2004), and intentions to perform health behaviors (Sirois 2004).

The general question addressed by this study is, how do those who give more consideration to long-term consequences differ from those who give more consideration to short-term consequences with regard to reactions and beliefs about wildfires and management? Are those who give more consideration to long-term consequences more likely to be concerned about wildfires, for example? Are they more likely to engage in risk reduction activities (such as taking defensible space measures around their homes)? Do they have particular preferences concerning how they receive information about wildfires?

Salient Value Similarity and Trust of the USDA Forest Service

Trust, the psychological willingness to rely on others or cooperate because of positive expectations of another person's intentions or behavior (Rousseau et al. 1998), is an important component of public responses to a broad range of risks (Siegrist 2000, Siegrist et al. 2000). Trust seems to be issue and situation specific (Kneeshaw et al. 2004, Langer 2002, Winter et al. 2004). An agency might be more trusted to manage one particular risk than another risk. Trust has been documented as an essential component of effective communication surrounding risk management (Covello et al. 1986, Freudenberg and Rursch 1994, Johnson 2004, Slovic 2000). Those who trust the source of a communication are more likely to believe the communicated message and more likely to accept initiatives designed to address that risk, including actions they must take themselves. In addition, trust has been found to be an important component of public responses to wildfire management (e.g., Liljeblad and Borrie 2006; Shindler et al. 2004; Winter et al. 2004; Winter and Cvetkovich 2004a, 2004b).

Those who trust the source of a communication are more likely to believe the communicated message and more likely to accept initiatives designed to address that risk.

Among studies examining trust related to forest-management issues, those most closely related to this study examined the interactions between salient values similarity and trust. In these studies, salient values similarity was a significant predictor of public trust in the Forest Service to address a number of natural resource management issues including a proposed program of research (Cvetkovich et al. 1995), a recreation fee demonstration program (Winter et al. 1999), and acceptance of approaches to manage threatened and endangered species (Cvetkovich and Winter 1998, 2003; Winter and Knap 2001). Other significant influences that have been explored in studies of trust related to forest-management issues include community of interest and place, ethnicity, gender, concern about the management issue in question, and knowledge about the target topic (Winter and Cvetkovich 2007).

In one study (Cvetkovich and Winter 2003), participants repeatedly raised the issues of the perceived consistency between Forest Service actions and similar salient values. From this we built a pair of items and tested them with publics regarding issues of endangered species management (Cvetkovich and Winter 2003) and fire management (Cvetkovich and Winter 2007). Perceived consistency between similar salient values and Forest Service actions, and justification of perceived inconsistency were instrumental in further understanding patterns of trust and distrust among publics. These findings are outlined in greater detail elsewhere (Cvetkovich and Winter 2004). The previous study of attitudes toward fire and fire management (Cvetkovich and Winter 2007) involved random samples of residents residing in four Southwestern States, including those with little direct experience with fire. In this study, we attempted to confirm that consistency and justification of inconsistency contribute to trust of Forest Service fire management, this time among communities known to have direct personal experiences with fire. Some of those receiving results from our four-state study asked us to report on residents' views that were known to be directly affected by fire risk. This study addresses their request.

METHODS

Participants

Residents and homeowners (n = 89) in fire-prone communities surrounded by the San Bernardino National Forest[1] participated in this study (table 1). We sought to obtain a purposive sample, rather than a random sample, of selected community areas, using key-contact and snowball approaches linked to the preexisting groups. Participants were invited through fire safe councils, local announcements in newspapers and radio stations, an e-mail tree through a forest district focusing on partnerships, and personal phone calls from the investigators. The majority (57.3 percent) of participants were male, White (92.1 percent), 55 years of age or older (68.6 percent), with at least some college education (85.3 percent, with 30.3 percent reporting some graduate study). A little more than one-fourth of participants (25.8 percent) had total household incomes of $49,999 or less. Other participants reported incomes from $50,000 to $74,999 (13.5 percent), or $75,000 or greater (42.7 percent).

A note on participation

Some community residents did not participate because of road closures or weather-related concerns (we had an unusual series of snowstorms, icy roads, and fog during the study period that came late in the season and kept many away because of safety concerns). A few residents expressed the feeling of being "meetinged out," considering what they judged to be an extensive number of meetings related to fire issues within their communities. Some told us they would only attend if a direct tangible benefit from their participation could be identified in advance of the meeting, while others expressed the feeling that they were waiting for action on prior meetings already held about fire management issues of concern to them before they would participate in more. Others told us they felt there was not adequate notice about our meetings. This was in spite of the radio and newspaper announcements, including media Web sites, as well as e-mail notices and telephone calls from the researchers or through fire safe councils. Identifying the most effective communication networks, including those that are community based, was an important part of our research effort, and we had only partial success. On one forest district, many of our contacts came through an e-mail tree derived from various partnership and collaborative efforts. This proved an invaluable resource to us, and the direct contact from someone residents knew in the Forest Service helped pave the way. We found that a number of routes and contacts were necessary. These routes differed greatly and in some ways reflected the unique nature of the communities we tried to reach.

Table 1. Schedule of focus groups and number of participants

Date	Location	Number of participants
March 18th	Angelus Oaks	12
March 21st	Forest Falls	8
March 22nd	Lake Arrowhead	3
March 23rd	Crestline	14
March 25th	Big Bear	7
March 26th	Wrightwood	12
March 30th	Idyllwild	17
March 31st	Lake Arrowhead	9
April 1st	Crestline	4
April 1st	Forest Falls	3

Survey Instrument

A self-administered questionnaire (app. A) created for this study included a number of Likert-type items focusing on three sets of reactions to and beliefs about wildfires and wildfire management:

1. Personal stress-related consequences of directly experiencing wildfires and living in communities threatened by wildfires (a series of yes/no items, adapted from the Impact of Event Scale-Revised, cited in Weiss and Marmar 1996) concern about the risk of wildfires (concern held by self and judged concern of other residents), and assessments of level of knowledge about wildfire management (self, residents, and Forest Service).
2. The perceived level of responsibility for wildfire prevention of various parties, effectiveness of risk reduction among responsible parties, personal participation in fire management activities (a series of yes/no items), and perceived barriers to effective fire management.
3. Views about preferred ways of receiving communication and education about wildfires and management.

In addition, a 12-item measure of consideration of future consequences created by Strathman et al. (1994) was included on the questionnaire. Measures of trust, salient values similarity, value consistency of actions, and justification of inconsistencies were adapted from earlier research reviewed in Cvetkovich and Winter (2007).

Focus Group Protocol

Participants were led through a series of discussion topics regarding fire and fire management on the San Bernardino National Forest (app. B). These items included objectives for fire management, concerns in fire management, alternatives to accomplish fire management objectives, shared values and trust in Forest Service fire management, and preferences for receiving communication and education.

Participants completed the self-administered questionnaire and then were led through the discussion topics.

Procedure

Each session lasted approximately 11/2 hours and started with a statement of purpose of the study, the voluntary nature of responses, importance of respect of other views in the discussion, and ability to opt out of any questions that made the participant uncomfortable. Participants completed the self-administered questionnaire and then were led through the discussion topics. Each discussion was audio taped; a notetaker recorded key comments and concepts to help anchor the transcription of audio records. Notes and surveys were matched through assigned identification numbers, allowing comparison between written and verbal responses.[2] Ten sessions were conducted over a 3-week period.

RESULTS

Reactions and Beliefs about Wildfires and Management

Personal experiences with fire

Participants reported a number of personal experiences with fire during their lifetimes. The vast majority had encountered wildland-fire related events such as seeing a fire (96.6 percent), smelling smoke (89.9 percent), and experiencing road closure (87.6 percent). Additional experiences shared by the majority included evacuation from their homes (69.7 percent), having power shut off to reduce fire risk (65.2 percent), and having a prescribed burn near their homes (62.9 percent). Less common were loss or damage to personal property of family, friend, or close neighbor (44.9 percent); personal loss or damage to property (15.7 percent); health problems or discomfort (15.7 percent); personal injury (5.6 percent); and family, friend, or neighbor suffering injuries (5.6 percent). Reported health problems were primarily smoke-related. On average, 6 of the 11 personal experiences listed were reported by each respondent. In judging the direct, personal impact of fire, a majority of participants (61.8 percent) selected a 6, 7, or 8 on the 8-point impact scale (1 = no impact, 8 = extensive impact) with only about one-tenth (9 percent) selecting 1, "no impact." In sum, the vast majority had personally experienced a number of fire-related impacts, and fire was judged to have a direct personal impact on most of the respondents' lives (figure 1).

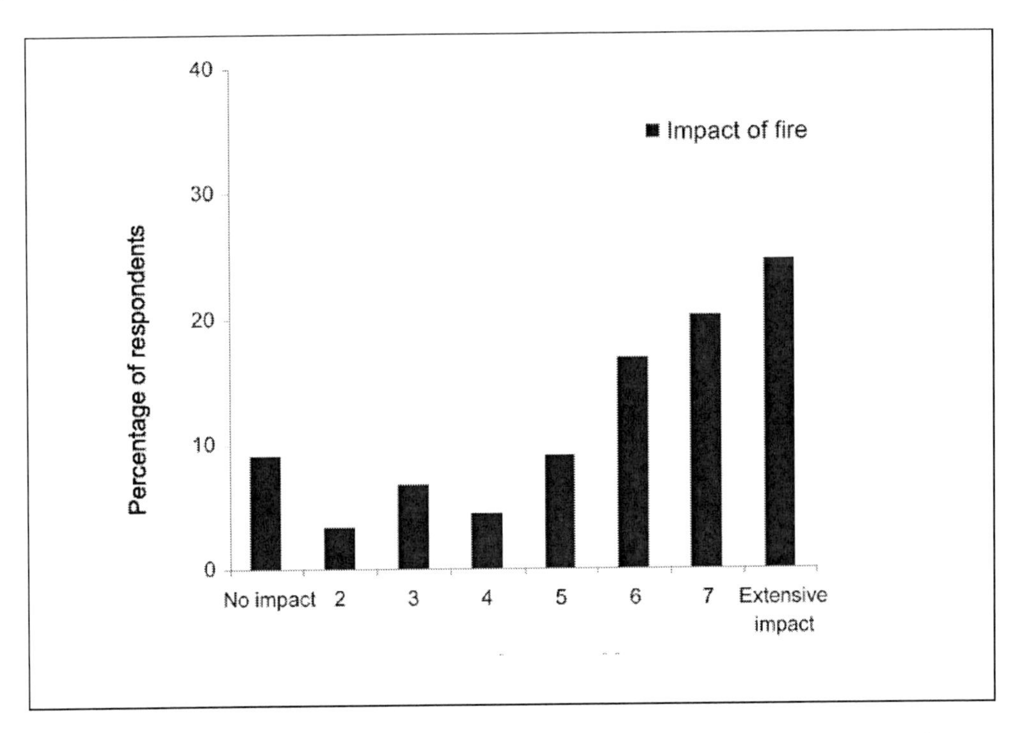

Figure 1. Degree of impact that fire on the San Bernardino National Forest has had on respondents

Personal consequences of fire and fire risk

The impact of living in a fire-prone ecosystem was examined through stress-related effects. Almost one-third of participants had not experienced any of the 21 listed possible difficulties resulting from wildland fire risk (the modal response was 1) in the past 7 days. Those reporting a greater number of fire-related experiences rated themselves as having more fire-related difficulties (as defined by the Weiss and Marmar scale, r = 0.37, $p < 0.001$, n = 83). Slightly more than one-third (38.2 percent) agreed that "I avoided letting myself get upset when I thought about it or was reminded of it," and almost one-third (29.2 percent) reported "any reminder brought back feelings about it," as well as "I felt watchful or on guard." About one-fourth (25.8 percent) reported that "other things kept making me think about it," and that "pictures about it popped into my mind" (24.7 percent). About one-fifth (18.0 percent) thought about it when they didn't mean to. Approximately one-tenth of our respondents reported "I had waves of strong feelings about it" (13.5 percent), "I tried not to think about it" (11.2 percent), "I felt irritable and angry" (9.0 percent), and feeling like they were back in a time when there was no fire (9.0 percent). Reporting of physical symptoms (sweating, trouble breathing, or nausea) was rare (only 3.4 percent). However, more than one-third (41.0 percent) indicated that more than one difficulty was experienced within the past 7 days. These results do not indicate major disruptions to everyday functioning. They do suggest that there is a continuing psychological impact from fire and fire risk even a few years after the last major fire in the local area.

There is a continuing psychological impact from fire and fire risk even a few years after the last major fire in the local area.

Concern about fire risk and knowledge about fire management

Participants rated their personal concern about fires and fire risks as high ($M = 7.43$, SD = 0.99; 1 = not at all concerned, 8 = very concerned; figure 2). Other community residents were also perceived as concerned, but not as concerned as self ($M = 6.70$, SD = 1.38; $F_{(1, 85)} = 23.08$, $p < 0.01$; figure 2).

Personal knowledge of what effective fire management should be done was rated as high ($M = 6.13$, SD = 1.60; 1 = not very knowledgeable, 8 = very knowledgeable; figure 3), but lower than ratings assigned to knowledge of fire management held by the Forest Service ($M = 6.86$, SD = 1.32; $F_{(1, 85)} = 13.70$, $p < 0.001$, figure 3). The knowledge of other community residents ($M = 3.92$, SD = 1.48; figure 3) was rated as lower than both the level of one's own knowledge ($F_{(1, 85)} = 139.71$, $p < 0.001$) and that of the Forest Service ($F_{(1, 85)} = 287.45$, $p < 0.001$).

Participants who rated their own concern and knowledge as high also tended to rate the concern and knowledge of other community members as high (r = 0.31, $p < 0.004$, n = 86 and r = 0.36, $p = 0.001$, n = 88, respectively). Judgments of Forest Service knowledge were not related to judgment of own knowledge (r = 0.19, $p = 0.08$, n = 88), but those who rated the knowledge of other citizens as high also rated the Forest Service's level of knowledge as high (r = 0.32, $p = 0.02$, n = 88). In sum, participants characterized their own concern and knowledge as higher than but similar to that of other community members, particularly if other community members shared their values. The Forest Service was characterized as more knowledgeable than either self or other community members.

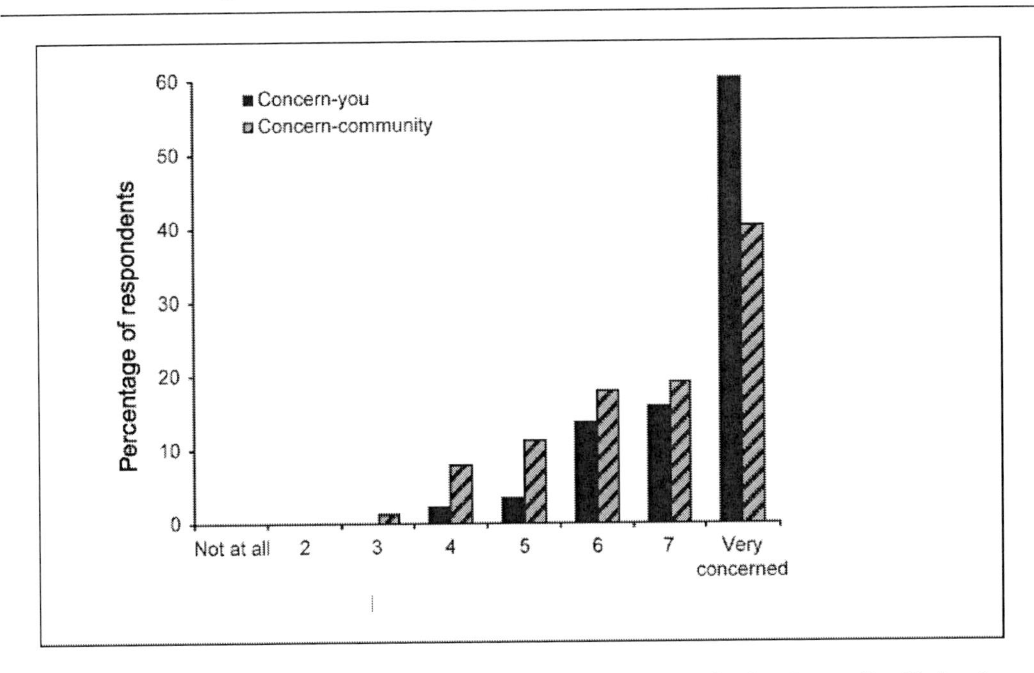

Figure 2. Ratings of concern of self and other community residents on the San Bernardino National Forest.

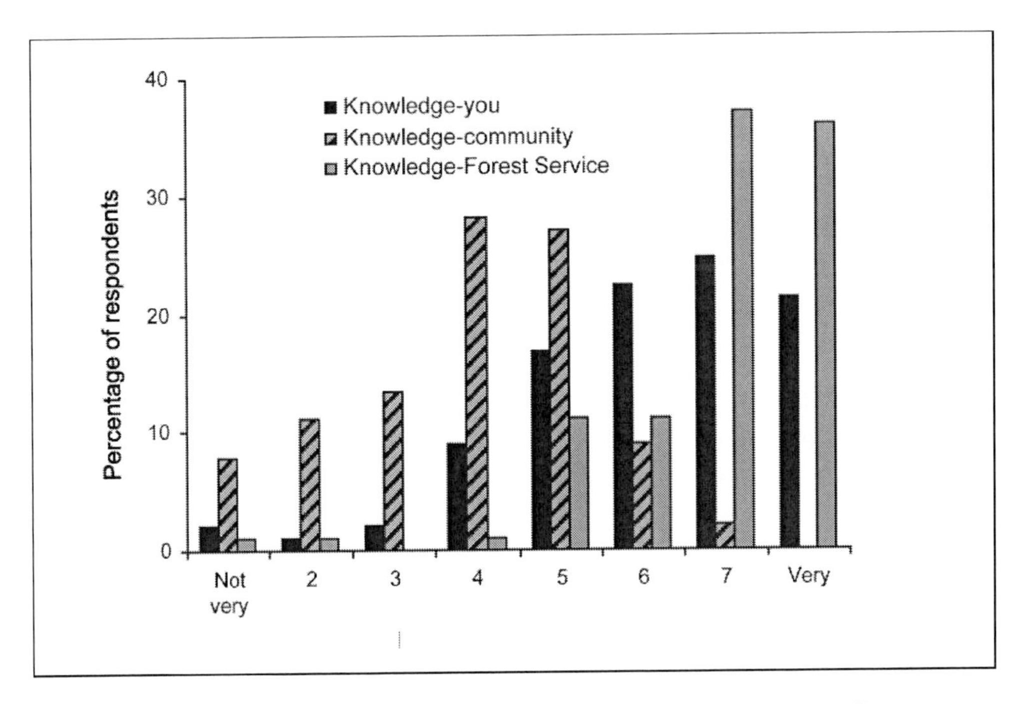

Figure 3. Ratings of knowledge of each party about what should be done for effective fire manage

Table 2. Number of points (out of 100) of responsibility in reducing the risk of wildland fires and grade on how well each party has done in the past 12 months in reducing the risk of wildland fires on the San Bernardino Mountains

	Participants assigning 1 or more points	Median responsibility	Mean responsibility	Std. Dev.	Range of points	Median grade
		Percent				
USDA Forest Service	98	20	18.68	11.89	5-80	B
California Department of Forestry	91	10	14.27	10.01	2-50	B
Local fire departments	88	10	11.51	8.32	5-20	A
Me and the people who live with me	88	10	11.81	12.39	1-80	B
My local community	88	10	10.79	9.06	3-50	C
Visitors and tourists	76	5	5.58	6.47	1-30	D
Federal legislators and representatives	75	10	8.79	8.60	1-40	C
State legislators and representatives	70	5	6.70	6.23	1-25	C
Scientists and researchers	61	5	4.21	4.35	1-20	C
Local business owners	61	5	5.00	3.94	1-13	C
Other	9	10	12.73	15.23	5-50	F

Perceived Responsibility for Fire Management

Perceived responsibility for fire management was assessed by asking participants to distribute 100 points among 10 potentially responsible parties. An “other” option was provided so that respondents could add parties to the list. Respondents could leave point assignments blank or enter “0” for no responsibility for reduction of fire risk. A followup question asked respondents to assign a grade to any party they had assigned points to. The grade was based on how well in the past 12 months the party had reduced the risk of wildland fires on the San Bernardino Mountains.

As shown in table 2, of the listed parties, the Forest Service was judged by the largest number of participants (over 80 percent) to have at least some responsibility and on average (both mean and median) was assigned the largest responsibility, around 20 points. Ratings of responsibility suggest that after the Forest Service, the California Department of Forestry (now called Cal Fire) is viewed as having a primary responsibility. It is also interesting that “me and the people who live with me” received fairly high responsibility ratings.

The Forest Service received a median grade of B for its fire-reduction efforts over the last year. Six of the listed parties were assigned a median responsibility of around 10 percent. One of these, “my local fire department,” was assigned a grade of A. The California Department of Forestry and “me and the people who live with me,” received grades of B. The remaining two, “my local community” and “federal legislators and representatives,” received a grade of C.

Only nine participants identified a responsible “other” in addition to those listed. Identified “others” included “lawsuits and regulations,” “local/county planning and regulations," "environmental groups," and "fire safe councils." These self- identified parties were assessed as having relatively high responsibility of around 10 points. Opting to identify a party not listed seemed to be prompted by a desire to identify those perceived as not doing a good job. Eight of the nine reported "others" were graded as deserving either a D or an F.

The remaining four listed parties were assigned an average of five points of responsibility. Three of these received an average grade of C. The fourth, "Visitors and tourists," received a grade of D.

Goals of Fire Management

During the focus group, discussion participants were asked, "What objectives for fire management are critical for this forest? Specifically, what should fire management accomplish on this forest... what should it do?" Several of the 143 responses identified the major goal of fire management to be the reduction of tree density (18.2 percent of responses), fuel removal (7.7 percent), and/or prescribed burns (3.5 percent) in order to establish and maintain a healthy "natural" forest (10.5 percent). Some concern was expressed about both the risks of prescribed burns and the need for communicating to residents when they were occurring (2.8 percent) and having the protection of people and property as the major goal of fire management (2.8 percent). Education about wildland fire and management was also expressed as an important goal (9.1 percent). Some participants were particularly concerned that not enough was being done focused on communicating to and educating nonresident tourists and backcountry users (2.1 percent). Appropriately timed closures of high risk areas were also identified as a technique for reaching fire management goals (2.1 percent). The planning and control of residential and other human development was also identified as an important component of effective fire management (5.6 percent). Three aspects of management during fire events were mentioned. These were the need for communication between officials and the public (5.6 percent), getting up-to-date news about the status of evacuation routes (5.6 percent), and coordination between different agencies (3.5 percent).

Table 3. Reported barriers to personal fire management activities

Barrier	Checked "yes"
	Percent
Inadequate financial	22.6
Own physical limitations	22.6
Don't want to change the landscape	21.8
Don't want to change my roof or other built structures	20.7
Not worried about fire	19.8
Not sure what will work	14.3
Don't know who to call/hire	3.6

Fire Management Activities

A number of actions that could effectively reduce fire risk were reported. Most people had read about what could be done to protect their homes from wildland fires (97.8 percent), had implemented defensible space around their property (94.4 percent), and had attended a public meeting about wildland fire (93.3 percent).[3] A majority had also reduced flammable vegetation on their property because they were required to do it (75.3 percent), worked with a community effort focused on fire protection (75.3 percent), made inquiries of the local fire safe council or volunteers on how to reduce fire risk (73.0 percent), made inquires of the local fire department on how to reduce fire risk (64.0 percent), and made inquiries of the local forest ranger (56.2 percent). A little over a third had changed the structure of their home to reduce risk (38.2 percent) and/or worked on a wildland fire suppression effort either in a paid or volunteer position (38.2 percent). Others had volunteered through various efforts or had worked through a fire safe council. An overall judgment yielded a moderately high evaluation of the effectiveness of these actions ($M = 6.01$, SD = 1.55, n = 85, median = 6; 1 = not at all effective, 8 = extremely effective).

Most people had read about what could be done to protect their homes from wildland fires, had implemented defensible space around their property, and had attended a public meeting about wildland fire.

Barriers to Personal Action

From about 4 percent to a little over 20 percent of participants indicated that their own fire reduction effort had been hindered because of one of the listed barriers (table 3). "Inadequate financial resources," "Own physical limitations," and "Don't want to change the landscape" were the most frequently reported barriers, followed closely by "Don't want to change my roof or other built structures" and "Not worried about fire risk." "Not sure what will work," and "Don't know who to call/hire" were the least frequently reported barriers.

Participants with lower incomes were more likely than those with higher incomes to report the barriers of "Own physical limitations" ($r = -0.41$, $p < 0.001$, n = 71), "Not sure what will work" ($r = -0.35$, $p < 0.001$, n = 85), "Not worried" ($r = -0.29$, $p < 0.02$, n = 71), "Inadequate financial resources" ($r = -0.27$, $p < 0.03$, n = 70), and "Don't want to change the landscape" ($r = -0.24$, $p < 0.05$, n = 72). Gender, age, and education level were not correlated with reported barriers.

Barriers to Others' Actions

A sizable number of participants concluded that effective reduction of fire risk has been hindered because at least one of the other involved parties had not done its part. About one-half (50.6 percent) believe that their neighbors have not done their part; about one-third (29.2 percent) believe public agencies have not done their part; and about one-fifth (22.5 percent) believe the Forest Service has not done its part. Those who reported that their neighbors have not done their part were also likely to cite the inactivity of public agencies (r = 0.47, $p < 0.001$, n = 85) and the Forest Service (r = 0.36, $p < 0.001$, n = 82) as barriers to reducing fire risk. Other barriers to effective risk reduction added by respondents in an open-ended question identified land use policies, growth and housing, community restrictions on removal of trees and vegetation, a lack of coordination between agencies, and environmentalists.

Communication and Education

Participants had many views on approaches to communication, collaboration, and education about fire management. The most preferred sources of information were public meetings the Forest Service leads so the community can ask questions (88.8 percent) and community meetings (84.3 percent). Other information sources preferred included a Web site (79.8 percent), brochures and pamphlets available on request (77.5 percent), articles in the local paper (77.5 percent), an e-mail tree sent by Forest Service representative and forwarded by fire safe council volunteers (75.3 percent), local television/radio spots put on by local Forest Service ranger (64.0 percent), and information and displays at Forest Service visitor center (60.7 percent). Additional suggestions included e-mails directly from the Forest Service, signs, a hotline or number residents could call to speak directly with someone knowledgeable, and messages on community bulletin boards. Flyers and newsletters left on residence doors were also brought up as a means of "getting the word out." It should be noted that the strong support for community meetings and direct engagement with the Forest Service was expressed by participants who themselves had come to participate in a meeting. As noted earlier, some residents expressed clear hesitation to participate in yet another meeting about fire.

The most preferred sources of information were public meetings the Forest Service leads so the community can ask questions, and community meetings.

Consideration of Future Consequences

The 12-item consideration of future consequences (CFC) scale (table 4) used to examine future orientation among respondents showed a comparatively high future orientation (alpha = 0.522).

The average CFC score for participants ($M = 4.15$, SD = 0.50, n = 89) was slightly higher (suggesting participants are somewhat more oriented to the future) than that reported in earlier research on college students (Petrocelli 2003). The vast majority (94.4 percent) had a score of either 4 or 5 (CFC was either "somewhat" or "extremely" characteristic). A higher CFC score was correlated to being more educated (r = 0.39, n = 89, $p < 0.01$) but not to gender (r = 0.15, n = 89, $p > 0.05$), age (r = -0.05, n = 88, $p > -0.05$), years living in current home (r = -0.08, n = 87, $p > 0.05$), or years living in the San Bernardino National Forest (r = -0.04, n = 88, $p > 0.05$).

Consideration of future consequences scores were correlated to only a few measures of reactions to and beliefs about wildfires and wildfire management. Participants who indicated that they gave more consideration to future consequences more often reported that they had inquired at their local fire department about how to reduce fire risk (r = 0.23, $p < 0.05$, n = 87). They were also likely to report that they found themselves acting or feeling as though they were back in a time where there was a fire (r = 0.30, $p < 0.01$, n = 82), although they were less likely to feel irritable or angry (r = - 0.23, $p < 0.05$, n = 83).

Table 4. Responses to the consideration of future consequences (CFC) scale

CFC item	Extremely uncharacteristic	Somewhat uncharacteristic	Uncertain	Somewhat characteristic	Extremely characteristic	Don't know
I consider how things might be in the future, and try to influence those things with my day to day behavior.	0	2.2	2.2	48.3	47.2	0
Often I engage in a particular behavior in order to achieve outcomes that may not result for many years.	6.7	2.2	16.9	29.2	40.4	2.2
I only act to satisfy immediate concerns, figuring the future will take care of itself.	58.4	23.6	4.5	4.5	0	9.0
My behavior is only influenced by the immediate (i.e., a matter of days or weeks) outcomes of my actions.	56.2	19.1	6.7	5.6	1.1	10.1
My convenience is a big factor in the decisions I make or the actions I take.	30.3	25.8	18.0	19.1	2.2	3.4
I am willing to sacrifice my immediate happiness or well-being in order to achieve future outcomes.	4.5	9.0	7.9	55.1	23.6	0
I think it is important to take warnings about negative outcomes seriously even if the negative outcome will not occur for many years.	1.1	2.2	6.7	41.6	47.2	1.1
I think it is more important to perform a behavior with important distant consequences than a behavior with less-important immediate consequences.	1.1	4.5	24.7	39.3	28.1	1.1

Table 4. (Continued)

CFC item	Extremely uncharacteristic	Somewhat uncharacteristic	Uncertain	Somewhat characteristic	Extremely characteristic	Don't know
I generally ignore warnings about possible future problems because I think the problems will be resolved before they reach a crisis level.	58.4	18.0	9.0	6.7	2.2	4.5
I think that sacrificing now is usually unnecessary since future outcomes can be dealt with at a later time.	55.1	23.6	7.9	9.0	1.1	3.4
I only act to satisfy immediate concerns, figuring that I will take care of future problems that may occur at a later date.	59.6	18.0	7.9	6.7	2.2	4.5
Since my day to day work has specific outcomes, it is more important to me than behavior that has distant outcomes.	37.1	32.6	9.0	11.2	5.6	2.2

Trust, Salient Values, and Reasons for Relying on the Forest Service

Participants' ratings of the salient values similarity items indicated a perception of shared values ("shares values": $M = 6.61$, SD = 1.53, median = 7, n = 85; "similar goals": $M = 6.37$, SD = 1.75, median = 7, n = 84; "supports views": $M = 6.31$, SD = 1.56, median = 6, n = 81). Less than 5 percent of the participants provided ratings below the midrange on each of these items, indicating dissimilar values. Of the 44 comments made during the focus group discussions concerning the values shared with the Forest Service, 4.7 percent related to the preservation of life and property and nearly 26 percent (25.6 percent) related to protection of the forest and natural habitat.

Since ratings of "shares values," "same goals," and "supports views" were highly intercorrelated (r = 0.70 to 0.74, $p < 0.001$, n = 81 to 84), a single index of "Salient Value Similarity of the Forest Service" (SVS) was computed based on the mean of responses to these three questions. This scale showed high reliability (alpha = 0.88) and was used in subsequent analyses.

Participants were also asked to what extent they trust the Forest Service in their fire management efforts. Based on an 8-point scale (1 = I completely distrust the Forest Service, 8 = I completely trust the Forest Service), responses leaned towards trust ($M = 5.85$, SD = 1.68, median = 6, n = 86; figure 4), with the majority (64 percent) providing ratings of 6 through 8 on the trust item. Trust of the Forest Service was reflected in comments such as: "I think we're on the same page with the Forest Service"; "We are real fortunate, because the Forest

Service has been a good partner"; "We love where we live. We love looking at the beautiful mountains and everything up here. They [the Forest Service] want to maintain that and we want to maintain that too"; and "One thing the local Forest Service has in common with the community is to preserve the forest."

As expected based on the salient values similarity model described in the introduction, trust of the Forest Service was significantly correlated to SVS ($r = 0.69$, $p < 0.001$, $n = 81$). Participants who believed that the Forest Service shared their values, had the same goals, and supported their views concerning fire protection also trusted the Forest Service's fire management.

Participants who believed that the Forest Service shared their values, had the same goals, and supported their views concerning fire protection also trusted the Forest Service's fire management.

When asked the extent to which fellow community residents share their values about fire management, the average response was above the midpoint on the scale, indicating moderately shared values ($M = 5.58$, SD = 1.55, median = 6, $n = 81$). Participants who perceived that their values were shared by other community residents rated community residents concern about wildfire as high ($r = 0.42$, $p < 0.001$, $n = 80$).

Respondents were asked if they thought that people were generally not trustworthy (value of 1) or generally trustworthy (value of 8, with value scale ranging from 1 to 8). Trustworthiness of others was rated fairly high ($M = 6.45$, SD = 1.60, median = 7, $n = 88$). Those who found others to be trustworthy were also likely to indicate that they trusted the Forest Service ($r = 0.49$, $p < 0.00 1$, $n = 85$).

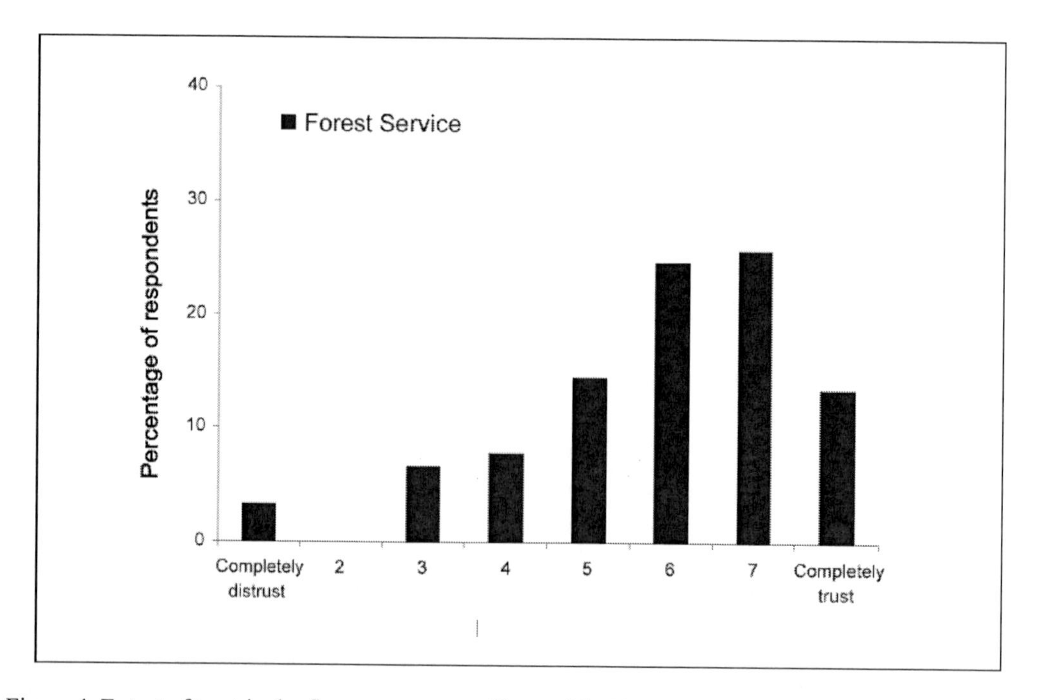

Figure 4. Extent of trust in the fire management efforts of the Forest Service.

Reasons for reliance

Participants indicated whether a series of items were reasons to rely on the Forest Service's fire management on the San Bernardino National Forest. A majority agreed or strongly agreed that the following were good reasons for relying on the Forest Service: "Procedures that ensure the Forest Service uses effective fire management" (67.4 percent), "Personal relationships I have with Forest Service personnel" (59.6 percent), and "The Forest Service's past record of fire management" (58.4 percent). A majority felt that the following were not reasons to rely on the Forest Service: "Media coverage of Forest Service fire management" (60.7 percent said this was not a reason), and "Congress holds the Forest Service accountable for its fire management" (52.8 percent said this was not a reason). Participants were almost equally divided on "Opportunities that I have to voice my views about fire management"; 38.2 percent said this was not a reason, 46.1 percent said it was a reason.

As shown in table 5, compared to those with less trust, those with more trust agreed that good reasons to rely on the Forest Service included "The Forest Service's past record of fire management," "Procedures that ensure the Forest Service uses effective fire management," and "Media coverage of Forest Service fire management." Compared to those who perceived less value similarity with the Forest Service, those who perceived more value similarity agreed that these three were good reasons to rely on the Forest Service as well as "Laws controlling the Forest Service's fire management," and "Personal relationships I have with Forest Service personnel."

However, step-wise multiple regression analyses of the reasons to rely on the Forest Service showed that of the seven reasons listed, only "Procedures that ensure the Forest Service uses effective fire management" was a significant predictor of level of trust (R2adj. (1, 76) = 0.13, $p < 0.001$) and level of SVS (R2adj. (1, 76) = 0.12, $p < 0.001$). Participants who were more trusting of the Forest Service and perceived greater value similarity agreed that procedures were a good reason to trust, those who were less trusting and perceived less value similarity tended to disagree.

Trust, concern, and knowledge about management

Individuals with high trust of the Forest Service judged both other residents (r = 0.30, n = 86, $p < 0.01$) and the Forest Service (r = 0.62, n = 86, $p < 0.01$) as having a high level of knowledge of fire management. Trust of the Forest Service was not significantly correlated to one's own level of concern about fire (r = -0.12, n = 86, $p > 0.05$), self-rated knowledge of fire management (r = -0.07, n = 86, $p > 0.05$), or other residents' level of concern (r = -0.13, n = 85, $p > 0.05$).

Trust, fire experiences, and personal consequences

Participants who had more fire-related experiences such as seeing a fire or knowing someone who lost property were less likely to trust the Forest Service (r = -.0293, $p < 0.01$, n = 83). Likewise, participants who reported more fire-related difficulties such as having waves of strong feelings or feeling watchful and on guard tended to trust the Forest Service less (r = 0.366, $p < 0.01$, n = 80).

Table 5. Correlations between trust, SVS, and reasons to rely on the Forest Service

Reason to rely on the Forest Service	Trust Forest Service		Forest Service shares values	
	r	n	r	n
Past record	0.37**	85	0.34**	84
Laws	0.30	84	0.32**	83
Personal relationships	0.14	84	0.30**	83
Procedures	0.33**	82	0.34**	81
Congress	0.09	81	0.16	80
Opportunity to voice my views	0.21	83	0.15	82
Media	0.31**	83	0.31**	82

** $p < 0.01$.

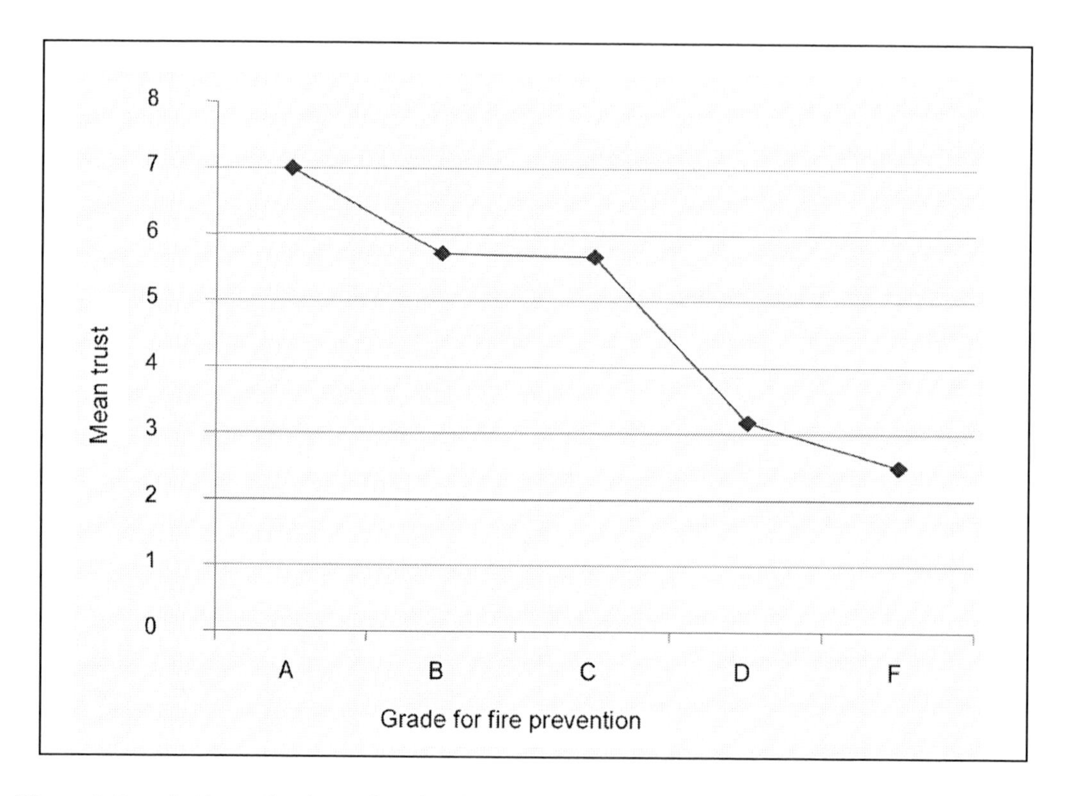

Figure 5. Trust in Forest Service and grade of Forest Service efforts in last year for fire prevention

Trust, responsibility, and evaluation of prevention effort

Points assigned to the Forest Service for level of responsibility for fire management were not correlated to level of trust (r = - 0.05, $p > 0.63$, n = 84). Analysis of variance of average trust ratings for grade assigned to effort to prevent fires found a significant main effect (F(4, 76) = 17.85, $p < 0.001$; figure 5) with those assigning higher grades indicating more trust. Scheffé analysis showed that participants who assigned a grade of A were significantly more trusting of the Forest Service than those assigning any other grade ($p < 0.02$). Those assigning

B and C did not differ in trust ($p = 1.0$), nor did those assigning grades of D and F ($p > 0.98$). The B and C graders were more trusting than D and F graders ($p > 0.02$).

Participants who were more trusting of the Forest Service gave higher grades to past efforts and perceived greater value similarity.

A stepwise multiple regression analysis showed that both grade of past fire prevention efforts (R2adj. (1, 78) = 0.44, $p < 0.001$) and level of SVS (R2adj. (1, 72) = 0.50, $p < 0.002$) were significant predictors of trust of the Forest Service. Participants who were more trusting of the Forest Service gave higher grades to past efforts and perceived greater value similarity.

Trust, activities, barriers, and communication

Neither the total number of actions taken to prevent wildfire ($r = -0.06$, $n = 81$, $p > 0.05$) nor the total number of perceived barriers to personal actions to reduce fire risk ($r = -0.03$, $n = 78$, $p > 0.05$) were significantly related to trust. Those who trusted the Forest Service less indicated that neighbors ($r = -0.22$, $n = 83$, $p < 0.05$) and public agencies ($r = -0.28$, $n = 82$, $p < 0.01$) had not done their part to prevent wildfires and were inclined to report that the Forest Service ($r = -0.25$, $n = 79$, $p < 0.05$) had not done its part. Preferences for particular sources of information about wildfires and fire management (e.g., local newspapers, e-mail trees, etc.) were not significantly correlated to trust.

Trust, value/action consistency and legitimacy

We asked participants to indicate how often the Forest Service makes decisions and takes actions consistent with their values, goals, and views. A small portion selected "never" (1.1 percent) or "rarely" (5.6 percent), and about one-fourth (25.8 percent) selected "sometimes." About one-third (33.7 percent) indicated Forest Service actions were usually consistent with their values, another fourth (24.7 percent) chose "almost always," and a few (2.2 percent) said Forest Service actions were always consistent with their values. Participants were then asked to respond to "If or when the Forest Service makes decisions or takes actions inconsistent with my values, goals, and views, the reasons for doing so are valid." A few disagreed with the statement (3.4 percent completely disagreed, and another 15.7 percent disagreed). Almost one-third (31.5 percent) neither agreed nor disagreed. Almost half agreed that an inconsistency between their own values and Forest Service actions was valid, when it occurred (39.3 percent agreed, 4.5 percent completely agreed). One participant expressed this balance between trust and valid reasons why the agency might not get things done, "I would trust one of them with my life. The only problem is red tape and money constraints." Another participant pointed to policy-related constraints, "What I am thinking is that the people in the Forest Service have the rulebook and are playing by the rulebook and the negligence comes with the change in policy. Maybe we need to have a more flexible policy. I trust the Forest Service people, but they are stuck with the policy and they need to figure a way to change policies."

Participants were categorized as either being above or below the midpoint of the response scales for salient values similarity, value consistency, and legitimacy of inconsistencies. This categorization identified four patterns of responses (table 6).[4]

Figure 6 shows mean trust of groups of participants with each of the patterns of categorical scores. The group of participants who rated salient values similarity low, value consistency low, and legitimacy of inconsistencies low (P1) were the lowest in mean trust ($M = 3.50$, SD = 2.57, n = 10). The group of participants who rated salient values similarity high, value consistency high, and legitimacy of inconsistencies high (P4) were the highest in mean trust ($M = 6.74$, SD = 1.12, n = 35). The other two patterns of the three ratings fell between these two extremes in trust of the Forest Service. Analysis of variance of mean trust showed a significant effect for pattern of ratings ($F(3, 74) = 21.44$, $p < 0.001$). A Scheffé test showed that patterns 1 and 2 were homogeneous and significantly different than the homogeneous subset of patterns 3 and 4 ($p = 0.05$).

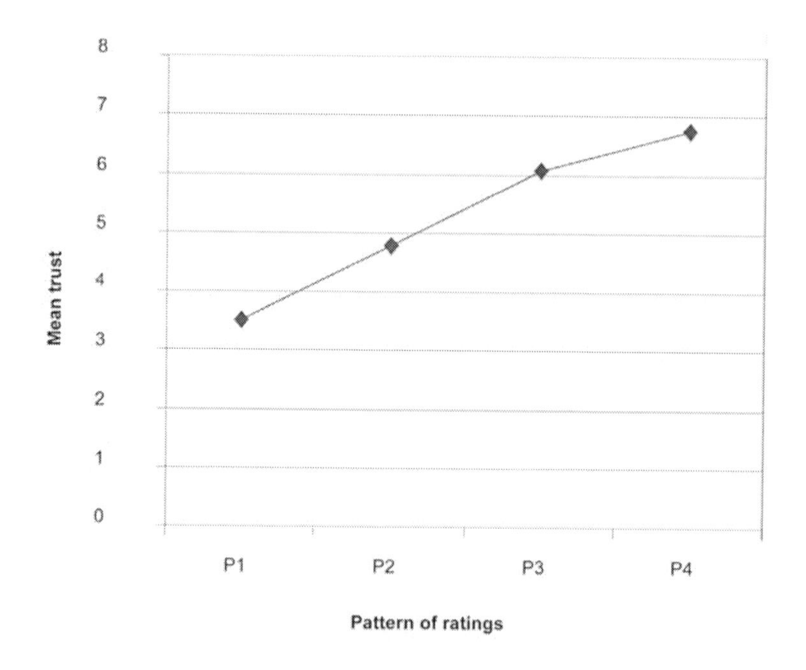

Figure 6. Trust in Forest Service and patterns of salient value similarity (SVS), value consistency, legitimacy of inconsistency (mean trust). P1: low SVS, low value consistency, low legitimacy; P2: high SVS, low value consistency, low legitimacy; P3: high SVS, high value consistency, low legitimacy; P4: high SVS, high value consistency, high legitimacy.

Table 6. Patterns of ratings of salient values similarity, value consistency, and legitimacy of inconsistencies

Pattern	Salient values similarity Low = 1-4 High = 5-8	Value consistency Low = 1-3 High = 4-6	Legitimacy of inconsistency Low = 1-3 High = 4-5	N
P1	Low	Low	Low	10
P2	High	Low	Low	18
P3	High	High	Low	15
P4	High	High	High	35
Missing				10

Discussion and Conclusion

Experiences in These Fire-Prone Communities

The majority of participants reported multiple fire-related experiences, although a minority had suffered personal injury or personal property loss. Almost half knew others who had suffered loss or damage. Comments about fire risk revealed that many took the risk of fire in stride, as part of living in the mountains. The one exception to this surrounded discussions about prescribed fire, where participants mentioned the risk of fires getting out of control, and the concern surrounding that management technique. A majority indicated that fire had an impact on them directly. The somewhat low rate of reporting stress-related experiences within the last 7 days probably reflected that the last fire event in the study area occurred over a year earlier. Another factor may have been the active role participants have taken in direct actions to reduce fire risk and to educate themselves about fire. This would be an interesting area for further research.

Both personal concern about fires and self-assessed knowledge of fire management were high. As participants lived in fireprone communities and had directly or indirectly experienced fires, these findings are not surprising. This high level of self-assessed knowledge does not seem unreasonable given that self was rated as lower in knowledge than the Forest Service and that other community members were judged to have similar, although somewhat lower, concern and level of knowledge. That almost all of the participants had reported taking personal actions to prevent wildland fires from harming their homes is in line with a high level of self-assessed knowledge. It would be useful for future research to validate both the self-assessed knowledge with an objective measure or test of knowledge about fire and the self-reported fire-risk actions with a direct objective method of assessment.

Responsibility and Performance

Participants were most likely to view agencies, especially the Forest Service, as holding a majority of responsibility for reduction of fire risk, with personal and community responsibility following closely. Agencies, including the Forest Service, personal households, and community were viewed as doing fairly well, although some respondents suggested the Forest Service and neighbors might not have always done their part in reducing fire risk. Although assigned little responsibility overall, tourists and visitors were viewed as doing poorly in reducing fire risk. Comments offered suggest that further limitations on tourists, including more limits on access or more limits on forms of use, were welcomed as additional measures to reduce fire risk.

Spontaneous comments indicated that participants considering the objectives of fire management were thinking about both fire prevention and firefighting. An important identified goal of fire prevention included the establishment and maintenance of healthy forests through various techniques such as fuel removal, reducing tree density, prescribed burns, planning and control of human development, and closure of high-risk areas. Important objectives of firefighting included coordination of different firefighting agencies, effective communication with the public, and making up-to-date information about the status of evacuation routes publicly available. Education was reported as an important objective of fire management by a number of the participants. Participants gave several useful suggestions for education and communication with regard to wildland fire management. Future research might investigate the degree of influence that education has on the actual practice of personal fire prevention activities, including the relative effectiveness of various educational approaches, the fit to differing communities, and the characteristics of those who seek education compared to those who do not.

Implications for Communication and Education

A majority of the participants supported public meetings with the Forest Service, and comments made clear the need to have an open forum where they could ask questions and receive answers from a knowledgeable source. Most of the methods of communication listed are already practiced within these communities to some degree or another, although some expressed the feeling that it had been a while since they had met with the Forest Service and they were starting to feel out of touch with what was going on. Others who did not attend the study sessions expressed a sense of overload on meetings. Clearly a variety of contacts needs to be practiced on an ongoing basis, and the use of community organizations and networks, including the fire safe councils, seems to be an effective vehicle to include. Although media were included in the means of contact, the local paper received more support than television or radio spots. A Web site for current and community-based information seemed to receive strong support. One community declined participation because they were waiting for the agency to act on commitments made in prior meetings. This demonstrates the importance of following up with community members after meetings and keeping them informed on an ongoing basis. Even efforts to meet commitments would probably be helpful to report. If barriers were met, those could also be reported, as it seemed participants understood that funding, policies, and other challenges could prevent the Forest Service from taking action.

This demonstrates the importance of following up with community members after meetings and keeping them informed on an ongoing basis.

Consideration of Future Consequences

Unlike the positive correlations with environment-related attitudes and behavior found by previous research, consideration of future consequences was not found to be strongly related to reactions to and beliefs about wildfires and wildfire management. The few significant correlations found out of the large number calculated could be due to chance. The failure to find significant relationships could be due to the restricted variability of the consideration of future consequences scores for this group of participants. It could also be associated with our limited number of participants; a strong statistical relationship would be needed to detect significance. It is too early to conclude that consideration of future consequences is not related to reported reactions to and beliefs about wildfires and wildfire management. In particular, we expect this measure to have some association with preventative measures that are aimed at reducing risk, especially those that require a more substantial investment of personal resources including time or money.

Trust, Values, Actions, and Risk Responses

Although perceived salient values and trust were significantly related to each other, consistency between perceived shared values and actions taken by the Forest Service seemed to be more influential in determinations of trust than were the shared values alone. This may have been due to the relatively high average rating of perceived value similarity, paired with low variability. Direct personal experiences with fire, and stressful impacts, were both negatively associated with trust. These results indicate the importance of considering other factors beyond salient values similarity in understanding trust. Those participants who perceived that the Forest Service shared their values but engaged in unjustified value-inconsistent actions reported an average level of trust statistically indistinguishable from participants who perceived that the Forest Service did not share their values.

In line with research regarding other organizations and other risks (Earle and Siegrist 2006), trust was significantly related to perceptions of the Forest Service's past record of effectiveness in reducing fire risk. Given the role of trust in acceptance of agency actions and communications, we expected to find a relationship between direct actions and trust. However, the number of actions taken had no relationship to trust. This has interesting implications for study of the relationship between trust and public response. Only those actions directly advocated through the Forest Service might be expected to be influenced by trust and perceived similar salient values. Reliance on procedures and personal relationships seemed to be a factor in deciding to rely on the Forest Service's fire management efforts. Past record of fire management seemed a bit less important, but was still held by a majority as a reason for reliance.

Trust is the psychological willingness to rely on others or cooperate because of positive expectations of another person's intentions or behavior. Trust is a moral- based evaluation of the character of other people. The correlations with SVS found by this study and other studies indicate that trust is the belief that Forest Service personnel can be relied on because of judged similarity between the citizen's currently active values and the values attributed to the Forest Service. Earle, Siegrist, and colleagues have distinguished confidence as a reason for reliance or cooperation separate from trust (Earle and Siegrist 2006; Earle et al. 2007; Siegrist et. al. 2003, 2005, 2007). Confidence is based on an evaluation of past performance. A perception of good performance leads to a confident feeling that uncertainty is low and that things are under control. Although trust and confidence are appraisals based on different information, the present results indicate that they may be interacting sources of reliance and cooperation. This study's results showed that evaluation of fire prevention efforts during the last year (measured by an assigned letter grade) and level of SVS were both significant predictors of trust of the Forest Service. Participants who were more trusting of the Forest Service both gave higher grades to past efforts and perceived greater value similarity. Also, of seven possible good reasons to rely on the Forest Service, only selection of "past record of fire management" was a significant predictor of both trust and SVS.

A particular kind of past performance, one involving evaluations of moral character, was examined in this study. Consistent with our earlier work (Cvetkovich and Winter 2003), patterns of increasing ratings of SVS, value consistency, and legitimacy of inconsistencies were associated with higher levels of trust.

Taken together, these findings regarding trust and confidence illustrate that forest managers should be aware that judgments of trust are made within a historical context of past performances. Sometimes past performance may be mostly evaluated relative to information that things are under control, as suggested by Earle and Siegrist's (2006) idea of confidence. At other times, past performance may be evaluated relative to moral judgments of shared value similarity, as suggested by the SVS model. It should be noted that for many citizens participating in this study, wildfire management is a high-concern issue. High-concern issues are very likely to invoke moral judgments related to shared values (Cvetkovich and Nakayachi 2007).

Elsewhere we have noted that in addition to information about value similarities and past record, willingness to rely on others and cooperate might be based on other relational assurances (Cvetkovich and Winter 2007). Examples of relational assurances understudied with regard to trust include laws, watchdog activities by nongovernmental agencies and the media, and established personal relationships between citizens and Forest Service personnel. There is a need for future work to investigate how different relational assurances affect trust and confidence.

Gaps and Where We Go from Here

Participants were fairly homogeneous and not representative of the overall populations within these forest communities. Although we made a concentrated effort to recruit seasonal residents, only a few actually participated. A past study sheds light on differences between seasonal and year-round residents of the San Bernardino mountain communities (Vogt and Nelson 2004). Some participants in our study suggested that seasonal residents and those leasing or renting their properties were less concerned, and less similar in values to the Forest Service than were the year-round community members. Additional studies of the perceptions of both seasonal and year-round residents, including how these groups view and are viewed by the Forest Service and other fire management agencies, would be of interest. The lack of relationship between personal actions taken and trust levels was somewhat surprising, although the relatively small sample size and little variance in trust may have suppressed any relationship between these two variables. The interest for meetings with and information from the Forest Service, and an interest in maintaining an ongoing dialogue were made clear. The need to report on actions taken, progress made, and barriers experienced by the Forest Service in its fire management efforts, was affirmed. These steps would assist the agency in continuing to develop trust and a positive basis for interaction in these communities, where individuals sometimes view themselves as very alone in their efforts to reduce risk.

This diversity of connections affirms the importance of knowing how to fit communication and education efforts to each community, and demonstrates the

significant challenge facing the ongoing commitment to reducing fire risk through individual, neighborhood, community, and multiagency engagement.

Throughout this study we found that communication approaches have to be varied and tailored to the unique characteristics of what we found to be unique "places" in what might be considered in some views as a predictably homogeneous area. These communities have very distinct connections to the Forest Service. Many community members are active in fire safe councils as well as other community groups and organizations. This diversity of connections affirms the importance of knowing how to fit communication and education efforts to each community, and demonstrates the significant challenge facing the ongoing commitment to reducing fire risk through individual, neighborhood, community, and multiagency engagement.

Acknowledgments

We thank the citizens of the involved communities who made this study possible by giving their time and effort to participate in the study and by helping in other ways. Additional thanks goes to Tricia Abbas, Ruth Wenstrom, Mike Dietrich, Laurie Rosenthal, and Jim Russell, all with the USDA Forest Service, who provided reviews and support in various ways throughout this study. Thanks also to the Inland Empire Fire Safe Council Alliance, especially Laura Dyberg, for their invaluable assistance in helping to arrange sessions and get the word out to members and residents about our study.

Metric Equivalents

1 acre = 0.405 hectare

References

Bonnicksen, T. M. (2003). *Written statement for the record. Field hearing on crisis on our national forests: reducing the threat of catastrophic wildfire to central Oregon communities and the surrounding environment.* Committee on Resources, U.S. House of Representatives, Redmond, Oregon. August 25, 2003. frwebgate.access.gpo.gov/cgi_bin/getdoc.cgi?dbname_108)house_hearings & docid_f:89089.pdf. p. 24-29. (December 9, 2006).

Collins, C. M. & Chambers, S. M. (2005). Psychological and situational influences on commuter-transport-mode choice. *Environment and Behavior. 37(5)*, 640–661.

Covello, V. T., Winterfeldt, D. V. & Slovic, P. (1986). Risk communication: a review of the literature. *Risk Abstracts.* 3(October), 171–182.

Cvetkovich, G. T. & Nakayachi, K. (2007). Trust in a high-concern risk controversy: a comparison of three concepts. *Journal of Risk Research. 10(2)*, 223–237.

Cvetkovich, G. T., Winter, P. L. (1998). *Community reactions to water quality problems in the Colville National Forest: final report.* Bellingham, WA: Western Institute for Social Research; Department of Psychology. On file with: Wildland Recreation and Urban Cultures Research Unit, Pacific Southwest Research Station, U.S. Department of Agriculture, Forest Service, 4955 Canyon Crest Drive, Riverside, CA 92507.

Cvetkovich, G. T. & Winter, P. L. (2003). Trust and social representations of the management of threatened and endangered species. *Environment & Behavior. 35(2)*, 286–307.

Cvetkovich G. T. & Winter, P. L. (2004). Seeing eye-to-eye on natural resource management: trust, value similarity, and action consistency/justification. In: Tierney, P.T.; Chavez, D.J., tech. coords. *Proceedings of the 4th social aspects and recreation research symposium.* San Francisco, CA: San Francisco State University: 46–50.

Cvetkovich, G. T. & Winter, P. L. (2007). The what, how, and when of social reliance and cooperative risk management. In: Siegrist, M.; Earle, T.C.; Gutscher, H., eds. *Trust in cooperative risk management: uncertainty and skepticism in the public mind.* London, United Kingdom: Earthscan: 187–209.

Cvetkovich, G. T., Winter P. L. & Earle T. C. (1995). *Everybody is talking about it: public participation in forest management.* Paper presented at the American Psychological Association convention. On file with: Wildland Recreation and Urban Cultures Research Unit, Pacific Southwest Research Station, U.S. Department of Agriculture, Forest Service, 4955 Canyon Crest Drive, Riverside, CA 92507.

Earle, T. C. & Siegrist, M. (2006). Morality information, performance information, and the distinction between trust and confidence. *Journal of Applied Social Psychology. 36(2)*, 383–416.

Earle, T. C., Siegrist, M. & Gutscher, H. (2007). Trust, risk perception and the TCC model of cooperation. In: Siegrist, M.; Earle, T.C.; Gutscher, H., eds. *Trust in cooperative risk management: uncertainty and skepticism in the public mind.* London, United Kingdom: Earthscan: 1–50.

Ebreo, A. & Vining, J. (2001). How similar are recycling and waste reduction? Future orientation and reasons for reducing waste as predictors of self-reported behavior. *Environment and Behavior. 33(3),* 424–448.

Freudenberg, W. R. & Rursch, J. A. (1994). The risks of "putting the numbers in context": a cautionary tale. *Risk Analysis. 14,* 949–958.

Inland Empire Fire Safe Alliance. (2006). Inland Empire communities at risk. http://www.fireinformation.com/IECommunitiesatRisk.html. (December 2, 2006).

Johnson, B. B. (2004). Risk comparisons, conflict, and risk acceptability claims. *Risk Analysis. 24,* 131–45.

Joireman, J. A. (1999). Additional evidence for validity of the consideration of future consequences scale in an academic setting. *Psychological Reports. 84(3, Pt. 2)*, 1171–1172.

Joireman, J. A., Van Lange, P. A. M. & Van Vugt, M. (2004). Who cares about the environmental impact of cars? Those with an eye toward the future. *Environment and Behavior. 36(2),* 187–206.

Joireman, J. A., Van Lange, P. A. M., Van Vugt, M., Wood, A., Leest, T. V. & Lambert, C. (2001). Structural solutions to social dilemmas: a field study on commuters' willingness to fund improvements in public transit. *Journal of Applied Social Psychology. 31(3),* 504–526.

Kneeshaw, K., Vaske, J. J., Bright, A. D. & Absher, J. D. (2004). Situational influences of acceptable wildland fire management actions. *Society and Natural Resources. 17*, 477–489.

Langer, G. (2002) (July/August). Trust in government: to do what? *Public Perspective*, 7–10.

Liljeblad, A., Borrie, W. T. (2006). Trust in wildland fire and fuel management decisions. *International Journal of Wilderness Research. 12(1)*, 39–43.

Molloy, T. (2004), April 30. Fire threat grows in California forest as trees continue to die. media. (December 2, 2006).

Orbell, S., Perugini, M. & Rakow, T. (2004). Individual differences in sensitivity to health communications: consideration of future consequences. *Health Psychology. 23(4)*, 388–396.

Petrocelli, J. V. (2003). Factor validation of the Consideration of Future Consequences Scale: evidence for a short version. *Journal of Social Psychology. 143(4)*, 405–413.

Rousseau, D. M., Sitkin, S. B., Burt, R. S. & Camerer, C. (1998). Not so different after all: a cross discipline view of trust. *Academy of Management Review. 23(3)*, 393–404.

Shindler, B., Brunson, M. W. & Cheek, K. A. (2004). Social acceptability in forest and range management. In: Manfredo, M.; Vaske, J.; Bruyere, B.; Field, D.; Brown, P., eds. *Society and natural resources: a summary of knowledge*. Jefferson, MO: Modern Litho Press, 147–158.

Siegrist, M. (2000). The influence of trust and perceptions of risk and benefits on the acceptance of gene technology. *Risk Analysis. 20(2)*, 19 5–203.

Siegrist, M., Cvetkovich, G. T. & Roth, C. (2000). Salient value similarity, social trust, and risk/benefit perception. *Risk Analysis. 20(3)*, 353–362.

Siegrist, M., Earle, T. C. & Gutscher, H. (2003). Test of a trust and confidence model in the applied context of electromagnetic field (EMF) risks. *Risk Analysis. 23(4)*, 705–716.

Siegrist, M., Earle, T. C. & Gutscher, H. (2005). Perception of risk: the influence of general trust, and general confidence. *Journal of Risk Research. 8(2)*, 145–15 6.

Siegrist, M., Gutscher, H. & Keller, C. (2007). Trust and confidence in crisis communication: three case studies. In: Siegrist, M.; Earle, T.C.; Gutscher, H., eds. *Trust in cooperative risk management: uncertainty and skepticism in the public mind.* London, United Kingdom: Earthscan: 267–286.

Sirois, F. M. (2004). Procrastination and intentions to perform health behaviors: the role of self-efficacy and the consideration of future consequences. *Personality and Individual Differences. 37(1)*, 115–128.

Slovic, P. (2000). Informing and educating the public about risk. In: Slovic, P., ed. *The perception of risk.* London, United Kingdom: Earthscan Publications: 182–198.

Strathman, A. J., Gleicher, F., Boninger, D. S. & Edwards, C. S. (1994). The consideration of future consequences: weighing immediate and distant outcomes of behavior. *Journal of Personality and Social Psychology. 66(4)*, 742–752.

United States Department of Agriculture Inspector General. (2006). *Forest Service large fire suppression costs*, Report No. 08601-44-SF. http://www.usda. gov/oig/webdocs/08601-44-SF.pdf. (December 8, 2006).

Vogt, C.A. & Nelson, C. (2004). Recreation and fire in the wildland-urban interface: a study of year-round and seasonal homeowners in residential areas nearby three national forests—San Bernardino National Forest, California; Grand Mesa, Uncompahgre and Gunnison, Colorado; and Apalachicola National Forest, Florida. Michigan State

University. Unpublished report. On file with: Wildland Recreation and Urban Cultures Research Unit, Pacific Southwest Research Station, U.S. Department of Agriculture, Forest Service, 4955 Canyon Crest Drive, Riverside, CA 92507.

Weiss, D. S. & Marmar, C. R. (1996). The impact of event scale-revised. In: Wilson, J.; Keane, T.M. eds. *Assessing psychological trauma and PTSD*. New York, NY: Guilford Press: 399–411.

Winter, G., Vogt, C. A. & McCaffery, S. (2004). Examining social trust in fuels management strategies. *Journal of Forestry. 102(6)*, 8–15.

Winter, P. L. & Cvetkovich, G. T. (2004a). *Is value correspondent action a necessary element of trust?* Paper presented at the 84th Western Psychological Association. On file with: Wildland Recreation and Urban Cultures Research Unit, Pacific Southwest Research Station, U.S. Department of Agriculture, Forest Service, 4955 Canyon Crest Drive, Riverside, CA 92507.

Winter, P. L. & Cvetkovich, G. T. (2004b). *The voices of trust, distrust, and neutrality: an examination of fire management opinions in three states*. Paper presented at the 10th international symposium on society and resource management. On file with: Wildland Recreation and Urban Cultures Research Unit, Pacific Southwest Research Station, U.S. Department of Agriculture, Forest Service, 4955 Canyon Crest Drive, Riverside, CA 92507.

Winter, P. L. & Cvetkovich, G. T. (2007). Diversity in southwesterners' views of Forest Service fire management. In: Kent, B.; Raish, C.; Martin, W., eds. *Wildfire risk: human perceptions and management implications*. Washington, DC: Resources for the Future Press: 156–170.

Winter, P. L. & Knap, N. (2001). *An exploration of recreation and management preferences related to threatened and endangered species: final report for the Angeles, Cleveland, Los Padres and San Bernardino National Forests*. Unpublished report. On file with: Wildland Recreation and Urban Cultures Research Unit, Pacific Southwest Research Station, U.S. Department of Agriculture, Forest Service, 4955 Canyon Crest Drive, Riverside, CA 92507.

Winter, P. L., Palucki, L. J. & Burkhardt, R. L. (1999). Anticipated responses to a fee program: the key is trust. *Journal of Leisure Research. 31(3)*, 207–226.

APPENDIX A: FIRE AND FIRE MANAGEMENT QUESTIONNAIRE

Public reporting burden for this information collection is estimated to average 15 minutes per response, with an additional 90 minutes to participate in the group discussion that will follow. This time estimate includes the time required for reviewing instructions, considering responses, completing the form and discussion, and reviewing your completed forms. Send comments regarding this burden estimate or any other aspect of this collection of information, including suggestions for reducing this burden, to Department of Agriculture, Forest Service, 1621 N. Kent Street, Room 800 RPE, Arlington, VA Attention: Clearance Officer; and to the Office of Management and Budget, Paperwork Reduction Project (OMB # 0596-0186), Washington, DC 20503.

The following questions focus on your views about fire and fire management in the San Bernardino National Forest.

Note: For this first set of questions we will ask you about fire management. When we ask about that we are referring to forest management techniques to reduce fire risk as well as fire management and suppression during an actual fire. Please circle one number from 1 to 8 indicating your response to each question. For any item that you are unable or do not wish to answer, circle the "D/K; N/A" ("don't know or no answer") option.

1. How concerned are you about fire National Forest?

0	1	2	3	4	5	6	7	8
D/K N/A	Not at all concerned							Very concerned

2. In your opinion, how concerned are San Bernardino National Forest community residents regarding fire and the risk of fire?

0	1	2	3	4	5	6	7	8
D/K N/A	Not at all concerned							Very concerned

3. How knowledgeable are you about what should be done for effective fire management on the San Bernardino National Forest?

0	1	2	3	4	5	6	7	8
D/K N/A	Not very knowledgeable							Very concerned

4. How knowledgeable do you think San -dents are about what should be Bernardino National Forest?

0	1	2	3	4	5	6	7	8
D/K N/A	Not very Knowledgeable							Very knowledgeable

5. How knowledgeable do you think the Forest Service is about what should be done for effective fire management on the San Bernardino National Forest?

0	1	2	3	4	5	6	7	8
D/K N/A	Not very Knowledgeable							Very knowledgeable

6. To what extent do you believe the USDA Forest Service (FS) *shares your values* about fire management?

0	1	2	3	4	5	6	7	8
D/K N/A	FS does not share my values							FS shares my values

7. To the extent that you understand them, *does the FS have the same goals,* for fire management as you do?

0	1	2	3	4	5	6	7	8
D/K N/A	FS has dissimilar goals							FS has similar goals

8. To what extent does the FS support your views about fire management?

0	1	2	3	4	5	6	7	8
D/K N/A	FS does not support my views							FS supports my views

9. To what extent do you trust the FS in their fire management efforts?

0	1	2	3	4	5	6	7	8
D/K N/A	I completely distrust the FS							I completely trust the FS

10. How often is the following true? "The FS makes decisions and takes actions consistent with my values, goals, and views."

0	1	2	3	4	5	6
D/K N/A	Never	Rarely	Sometimes	Usually	Almost always	always

11. How much do you agree or disagree with the following? "If or when the FS makes decisions or takes actions inconsistent with my values, goals, and views, the reasons for doing so are valid."

0	1	2	3	4	5
D/K	Completely	Disagree	NeitherAgree	agree	Completely
N/A	disagree		or disagree		agree

12. There are various reasons why individuals may or may not rely on the Forest Service's fire management on the San Bernardino. Please rate each of the items below, using the following scale:

3 = I strongly agree, this is a reason that I rely on the Forest Service
2 = I agree, this is a reason that I rely on the Forest Service
1 = This is **not** a reason that I rely on the Forest Service
0 = I have no opinion or am not sure

Reasons	Circle one response for each reason			
The Forest Service's past record of fire management	3	2	1	0
The laws controlling the Forest Service's fire management	3	2	1	0
Personal relationships I have with Forest Service personnel	3	2	1	0
Procedures that ensure the Forest Service uses effective fire management	3	2	1	0
Congress holds the Forest Service accountable for its fire management	3	2	1	0
Opportunities that I have to voice my views about fire management	3	2	1	0
Media coverage of Forest Service fire management	3	2	1	0

13. Fire management can accomplish multiple and varied objectives. In your opinion, what are the most important objectives for fire management on this forest?

14. Earlier, we asked about trust of the FS in fire management efforts. In general, do you find people to be:

0	1	2	3	4	5	6	7	8
D/K N/A	Generally not trustworthy							Generally trustworthy

15. Earlier, we asked you about shared values with the FS. We'd also like to know to what extent your fellow community residents share your values about fire management. Would you say that they:

0	1	2	3	4	5	6	7	8
D/K N/A	Do not share my values							Share my values

16. Which of these personal experiences with fire have you had during your lifetime?

Personal Experiences with Fire	Circle no or yes for each	
Saw a wildland	No	Yes
Experienced smoke from a wildland		
A prescribed burn occurred near my	No	Yes
Experienced road closure due to	No	Yes
Was evacuated from my home because of wildland fire or risk of fire		
Went without power, shut off to reduce fire risk	No	Yes
Lost or suffered damage to personal property due to a wildland fire *If yes, approximate value of loss -- $*	No	Yes
Family, friend, or close neighbor lost or suffered damage to personal property due to a wildland fire	No	Yes
Was injured by a wildland fire *If yes, please describe*	No	Yes
Family, friend, or neighbor was injured by a wildland fire *If yes, please describe*	No	Yes
Experienced health problems or discomfort caused by smoke from a wildland *If yes, please describe*	No	Yes

17. Please rate the degree of impact that fire on the San Bernardino National Forest has had on you directly.

0	1	2	3	4	5	6	7	8
D/K N/A	No impact							Extensive impact

18. Following is a series of questions focused on difficulties people sometimes have after stressful life events. Please indicate which if any of the following difficulties you have experienced during the PAST SEVEN DAYS with respect to the risk of wildland fire.

Difficulties	Circle no or yes for each	
Any reminder brought back feelings about it	No	Yes
I had trouble staying asleep		
Other things kept making me think about it	No	Yes
I felt irritable and angry	No	Yes
I avoided letting myself get upset when I thought about it or was reminded of it	No	Yes
I thought about it when I didn't mean to	No	Yes
I felt as if it hadn't happened or wasn't real	No	Yes
I stayed away from reminders about it	No	Yes

(Continued)

Difficulties	Circle no or yes for each	
Pictures about it popped into my mind	No	Yes
I was jumpy and easily startled	No	Yes
I tried not to think about it	No	Yes
I was aware that I still had a lot of feelings about it, but I didn't deal with them	No	Yes
My feelings about it were kind of numb	No	Yes
I found myself acting or feeling as though I was back at a time when there was a fire	No	Yes
I had trouble falling asleep	No	Yes
I had waves of strong feelings about it	No	Yes
I tried to remove it from my memory	No	Yes
I had trouble concentrating	No	Yes
Reminders of it caused me to have physical reactions, such as sweating, trouble breathing, nausea, or a pounding heart	No	Yes
I had dreams about it	No	Yes
I felt watchful or on guard	No	Yes
I have not experienced any of these difficulties	No	Yes

19. Assuming you have 100 points to characterize full responsibility for reduction of fire risk, please assign the number of points (using whole numbers only please) you think each party has in reducing the risk of wildland fires on the San Bernardino Mountains.

Party	Points
Federal legislators and representatives	
State legislators and representatives	
Scientists and researchers	
Local fire departments	
The U.S.D.A. Forest Service	
California Department of Forestry	
Local business owners	
Visitors and tourists	
My local community	
Me and the people who live with me	
Other (*please fill in*)	
TOTAL	100

20. Taking only those who you assigned points to (even if 1 or 2), please assign each a grade (from A for excellent through F for failing, avoiding pluses or minuses) on how you think they have done in the past 12 months in reducing the risk of wildland fires on the San Bernardino Mountains. If you did not assign points to someone listed, please circle "N/A."

Party		Grade				
Federal legislators and representatives	N/A	A	B	C	D	F
State legislators and representatives	N/A	A	B	C	D	F
Scientists and researchers	N/A	A	B	C	D	F
Local fire departments N/A		A	B	C	D	F
The U.S.D.A. Forest Service	N/A	A	B	C	D	F
California Department of Forestry	N/A	A	B	C	D	F
Local business owners N/A		A	B	C	D	F
Visitors and tourists	N/A	A	B	C	D	F
My local community	N/A	A	B	C	D	F
Me and the people who live with me	N/A	A	B	C	D	F
Other (*please fill in*)	N/A	A	B	C	D	F

21. Which of the following actions have you taken as a resident in the San Bernardino Mountains?

Action	Circle no or yes for each	
Read about home protection from wildland fires	No	Yes
Attended a public meeting about wildland fire	No	Yes
Implemented defensible space around my property	No	Yes
Removed flammable vegetation on my property because I was required to do it	No	Yes
Made inquiries of the local fire department how to reduce risk of property damage from wildland fire	No	Yes
Made inquiries of the local forest ranger how to reduce risk of property damage from wildland fire	No	Yes
Made inquiries of the local Fire Safe Council office or volunteer(s) on how to reduce risk of property damage from wildland fire	No	Yes
Changed structure of my home to reduce risk of property damage from wildland fire	No	Yes
Worked with community effort focused on fire protection	No	Yes
Worked on wildland fire suppression effort as part of paid or volunteer position	No	Yes
Other (please describe)	No	Yes

22. If you circled yes for any actions you took that are designed to reduce the risk of losing your home during a wildland fire (in item 21), how effective do you think these actions are?

0	1	2	3	4	5	6	7	8
D/K N/A	Not at all effective							Extremely effective

23. Sometimes there are barriers to effective reduction of fire risk. Among the possible barriers listed below, please circle no or yes to indicate if a barrier (or barriers) apply to reducing the risk of fire in the area immediately surrounding your property.

Barrier	Circle no or yes for each	
I don't have adequate financial resources	No	Yes
My own physical limitations	No	Yes
I don't know who to call/hire to help	No	Yes
I don't want to change the landscape	No	Yes
I don't want to change my roof or other built structures	No	Yes
I am not sure what will really work	No	Yes
I am not worried about fire risk	No	Yes
My neighbors have not done their part	No	Yes
Public agencies have not done their part	No	Yes
The Forest Service has not done its part	No	Yes
Other (*please describe*)	No	Yes

24. For each of the statements below, please indicate whether or not the statement is characteristic of you. If the statement is extremely uncharacteristic of you (not at all like you) please answer "1"; if the statement is extremely characteristic of you (very much like you) please answer "5" next to the question. And, of course, use the numbers in the middle if you fall between the extremes. Please keep the following scale in mind as you rate each of the statements below.

0	1	2	3	4	5
D/K N/A	Extremely uncharacteristic	Somewhat uncharacteristic	Uncertain	Somewhat characteristic	Extremely characteristic

Statement	Circle one answer for each					
I consider how things might be in the future, and try to influence those things with my day to day behavior.	0	1	2	3	4	5
Often I engage in a particular behavior in order to achieve outcomes that may not result for many years.	0	1	2	3	4	5
I only act to satisfy immediate concerns, figuring the future will take care of itself.	0	1	2	3	4	5
My behavior is only influenced by the immediate (i.e., a matter of days or weeks) outcomes of my actions.	0	1	2	3	4	5
My convenience is a big factor in the decisions I make or the actions I take.	0	1	2	3	4	5
I am willing to sacrifice my immediate happiness or well-being in order to achieve future outcomes.	0	1	2	3	4	5
I think it is important to take warnings about negative outcomes seriously even if the negative outcome will not occur for many years.	0	1	2	3	4	5

(Continued)

Statement	Circle one answer for each					
I think it is more important to perform a behavior with important distant consequences than a behavior with less-important immediate consequences.	0	1	2	3	4	5
I generally ignore warnings about possible future problems because I think the problems will be resolved before they reach crisis level.	0	1	2	3	4	5
I think that sacrificing now is usually unnecessary since future outcomes can be dealt with at a later time.	0	1	2	3	4	5
I only act to satisfy immediate concerns, figuring that I will take care of future problems that may occur at a later date.	0	1	2	3	4	5
Since my day to day work has specific outcomes, it is more important to me than behavior that has distant outcomes.	0	1	2	3	4	5

25. How would you like to receive information from the FS regarding fire and reduction of fire risk?

Source of Information	Your preference		
Articles in our local paper	No	Indifferent	Yes
Attendance at community meetings	No	Indifferent	Yes
Public meetings the FS leads so community can ask questions	No	Indifferent	Yes
Information and displays at FS visitor center	No	Indifferent	Yes
Brochures and pamphlets available on request	No	Indifferent	Yes
Web site	No	Indifferent	Yes
E-mail tree sent by a FS representative and forwarded by Fire Safe Council volunteers	No	Indifferent	Yes
Local television/radio spots, put on by local FS ranger	No	Indifferent	Yes
Other (*please fill in*)	No	Indifferent	Yes

26. Check the highest grade or year of school that you have completed and received credit for.

Highest Grade or Year of School	Check only one
Middle school or less	
High school degree (or G.E.D.)	
At least one year of college, trade, or vocational school	
Graduated college with a Bachelor's	
At least one year of graduate work equivalent	
Don't wish to answer	

27. Check the category that contains your age.

Age Group	Check only one
18 to 24	
25 to 34	
35 to 44	
45 to 54	
55 to 64	
65 or over	
Don't wish to answer	

28. Which of the following ethnic groups best describes you?

Ethnic Group	Check
Hispanic or Latino/a	
Not Hispanic or Latino/a	

29. Which of the following racial categories best describes you?

Racial Categories	Check one or more
American Indian or Alaska	
Asian	
Black or African American	
Native Hawaiian or Other Pacific	
White, Caucasian, or Euro American	
Another ethnic or racial group	
Don't wish	

30. Check the one category that best describes your household's total income for last year before taxes?

Income	Check only one
Under $5,000	
$5,000 to $9,999	
$10,000 to $14,999	
$15,000 to $24,999	
$25,000 to $34,999	
$35,000 to $49,999	
$50,000 to $74,999	
$75,000 to $99,999	
$100,000 or more	
Don't	

31. How many years have you lived in your current home?

Years	

32. How many years have you lived within the perimeter of the San Bernardino National Forest?

Years	

33. What is your ZIP code?

ZIP code _	

APPENDIX B: FOCUS GROUP PROTOCOL

Perceptions of Risk, Trust, Responsibility, and Management Preferences Among Fire-Prone Communities on the San Bernardino National Forest

Hello and welcome. I want to thank you for coming here today.
My name is and I am here with ___________ and ___________.

We will be talking together about the Forest Service, other agencies, and residents in your community regarding fire and fire management. We want your own views as a community member, and will not be expecting you to represent other's opinions or a particular group you belong to aside from yourself. I have a few questions for you, and mostly want to hear from you about what your thoughts are. This is an open discussion and we want to encourage each of you to share your ideas, whether you feel others in the room may have already expressed that idea, or a contradictory opinion to your own. Since we want to hear from each of you, we are asking that you give each other a chance to speak, and that you treat each other with respect. If you have a cell phone with you either turn it to silent mode or turn it off so that our discussion is not interrupted. We will not be attempting to reach consensus on any topic, or ask for any votes.

Your identity will be kept confidential, but we need to identify speakers with their comments, and to match those with the questionnaire. We are using the ID assigned to you and as placed on the notecard in front of you. Once the responses are matched up, your identity is kept separate from the databases we will create. Any contact information we collected to invite you to this meeting will not be stored with these data. If you have any questions or concerns about this please see me at the end of the group session so we can explain our procedures to keep your confidentiality secure.

We will be having our group discussion for the next hour and a half. We will not take any breaks, but if you want to get up and move around feel free to do so quietly. We are tape recording and making written summaries of our discussion. This is just for our use, so we don't miss any of your ideas. These tapes will not be heard by anyone else other than myself and my research coding team. The transcripts will not contain your names. We ask that you speak one at a time, so that the recordings are clear and we can track the discussion.

Most people find these focus group discussions enjoyable and informative. I want to acknowledge that you are of course, under no obligation to answer anything that you do not wish to and that your participation is voluntary so you may leave at any time.

To begin things, let's go around the table and introduce yourselves... some of you may already know each other. Just give your first name, and tell how long you have been a resident of this community. Also, please let us know if this community is your primary or secondary place of residence.

Objectives/Values/Concerns

A. There are numerous objectives that fire management can address on a national forest. In your opinion, what objectives for fire management are critical for this forest? Specifically, what should fire management accomplish on this forest... what should it do?
B. What values are linked to the objectives we just discussed?
C. Views on key concerns about fire risk and fire management
   What are your main concerns about fire risk and fire management? Which are most important to you? We are asking you to tell us what concerns you the most about fire, not to evaluate current effectiveness of management. Examples might include scenic beauty, property values, community safety, etc.
D. Now that we've listed your concerns, do we need to go back and revise our objectives? Did we miss something in our first round of objectives that we need to pull in now?

Alternatives

E. What alternative approaches are there for arriving at the objectives this group has listed? Let's try to keep in mind costs, and the desire to address multiple objectives if possible
F. What are the expected impacts and consequences of these various alternatives?
G. What are the risks inherent in each of these alternatives?
H. Which alternatives then, in light of everything we've discussed so far, can you support?
I. What concerns do you have about these alternatives?
J. Now, as we've listed these objectives together, individuals, groups, and agencies were identified. Let's go through these and examine your level of trust in each... What's your level of trust in the various parties you see necessary to address each alternative?

Values/Goals and Trust

K. In the survey we asked you to indicate the level of shared values and goals you hold similar to the FS. What are the most important values and goals that the FS shares with you?
L. What are the most important ones that the FS does not share with you?
M. Thinking back to your trust/distrust rating for the FS, what did you consider in making your rating? Tell me what experiences or information came to mind as you were answering this question. Was it personal interactions, media accounts, or something else?
N. Let's go back to the idea of shared values with the FS, and see if you can think of instances when the FS acted in ways inconsistent with those values. Can you think of examples of when that has happened?
O. And if that happened, were the reasons for inconsistency valid in your mind, that is, when talking about inconsistency, can you think of reasons why that might have occurred? What might be some valid reasons for inconsistency? What might be some instances where inconsistency was not valid?
P. On the questionnaire, we asked the extent to which certain factors, such as the Forest Service's past record on fire management, affect your reliance on their fire management. Other factors were laws, personal relationships, existing procedures, accountability to Congress, opportunities to voice your opinions, and media coverage. What were you thinking about when you answered these questions? Is there anything that has been most influential in your reliance on Forest Service fire management?

Information Needs and Mode of Receipt

Q. What information, if any, would you be interested in receiving or do you feel you need from the FS regarding fires and fire management on this forest?

End Notes

1 One community is located in a part of the San Bernardino National Forest managed by the Angeles National Forest.
2 Participation was completely anonymous. Although first names and brief introductions were shared for group facilitation purposes, these were not recorded and only participant identification numbers were used in the gathering and recording of data. Participants were informed in advance that their responses and comments would be handled in this manner, in order to facilitate openness and candor.
3 We expect this high number is characteristic of the intensive effort to reduce fire risk and to raise awareness of fire management efforts in the participating communities.
4 One individual who was high on SVS, low value consistency, and high on legitimacy of inconsistency was categorized into pattern 2.

In: Wildfires and Wildfire Management
Editor: Kian V. Medina
ISBN: 978-1-60876-009-1

Chapter 3

FOREST FIRE/WILDFIRE PROTECTION

Ross W. Gorte

SUMMARY

Congress continues to face questions about forestry practices, funding levels, and the federal role in wildland fire protection. Recent fire seasons have been, by most standards, among the worst in the past half century. National attention began to focus on wildfires when a prescribed burn in May 2000 escaped control and burned 239 homes in Los Alamos, NM. President Clinton responded by requesting a doubling of wildfire management funds, and Congress enacted much of this proposal in the FY2001 Interior appropriations act (P.L. 106-291). President Bush responded to the severe 2002 fires by proposing a Healthy Forests Initiative to reduce fuel loads by expediting review processes.

Many factors contribute to the threat of wildfire damages. Two major factors are the decline in forest and rangeland health and the expansion of residential areas into wildlands—the wildland-urban interface. Over the past century, aggressive wildfire suppression, as well as past grazing and logging practices, have altered many ecosystems, especially those where light, surface fires were frequent. Many areas now have unnaturally high fuel loads (e.g., dead trees and dense thickets) and an historically unnatural mix of plant species (e.g., exotic invaders).

Fuel treatments have been proposed to reduce the wildfire threats. Prescribed burning—setting fires under specified conditions—can reduce the fine fuels that spread wildfires, but can escape and become catastrophic wildfires, especially if fuel ladders (small trees and dense undergrowth) and wind spread the fire into the forest canopy. Commercial timber harvesting is often proposed, and can reduce heavy fuels and fuel ladders, but exacerbates the threat unless and until the slash (tree tops and limbs) is properly disposed of. Other mechanical treatments (e.g., precommercial thinning, pruning) can reduce fuel ladders, but also temporarily increase fuels on the ground. Treatments can often be more effective if combined (e.g., prescribed burning after thinning). However, some fuel treatments are very expensive, and the benefit of treatments for reducing wildfire threats depend on many factors.

It should also be recognized that, as long as biomass, drought, and high winds exist, catastrophic wildfires will occur. Only about 1% of wildfires become conflagrations, but which fires will "blow up" into crown wildfires is unpredictable. It seems likely that management practices and policies, including fuel treatments, affect the probability of such events. However, past experience with wildfires are of limited value for building predictive models, and research on fire behavior under various circumstances is difficult, at best. Thus, predictive tools for fire protection and control are often based on expert opinion and anecdotes, rather than on research evidence.

Individuals who choose to build homes in the urban-wildland interface face some risk of loss from wildfires, but can take steps to protect their homes. Federal, state, and local governments can and do assist by protecting their own lands, by providing financial and technical assistance, and by providing relief after the fire.

The spread of housing into forests and other wildlands,[1] combined with various ecosystem health problems, has substantially increased the risks to life and property from wildfire. Wildfires seem more common than in the 1960s and 1970s, with 2005, 2006, and 2007 being the most severe fire seasons since 1960.[2] National attention was focused on the problem by a fire that burned 239 houses in Los Alamos, NM, in May 2000. Issues for Congress include oversight of the agencies' fire management activities and other wildland management practices that have altered fuel loads over time; consideration of programs and processes for reducing fuel loads; and federal roles and responsibilities for wildfire protection and damages.

Many discussions of wildfire protection focus on the federal agencies that manage lands and receive funds to prepare for and control wildfires. The Forest Service (FS), in the Department of Agriculture, is the "big brother" among federal wildfire-fighting agencies. The FS is the oldest federal land management agency, created in 1905, with fire control as a principal purpose. The FS administers more land in the 48 coterminous states than any other federal agency, receives about two-thirds of federal fire funding, and created the symbol of fire prevention, Smokey Bear. The Department of the Interior (DOI) contains several land-managing agencies, including the Bureau of Land Management (BLM), National Park Service, Fish and Wildlife Service, and Bureau of Indian Affairs; DOI fire protection programs have been coordinated and funded through the BLM. Despite the substantial attention given to the FS and DOI agencies, the majority of wildlands are privately owned,[3] and states are responsible for fire protection for these lands, as well as for their own lands.

This chapter provides historical background on wildfires, and describes concerns about the *wildland-urban interface* and about forest and rangeland health.[4] The report discusses fuel management, fire control, and fire effects. The report then examines federal, state, and landowner roles and responsibilities in protecting lands and resources from wildfires, and concludes by discussing current issues for federal wildfire management.

Historical Background

Wildfire has existed in North America for millennia. Many fires were started by lightning, although Native Americans also used fire for various purposes. Wildfires were a problem for early settlers. Major forest fires occurred in New England and the Lake States in the late 1800s, largely fueled by the tree tops and limbs (slash) left after extensive logging. One particularly devastating fire, the Peshtigo, is commonly cited as the worst wildfire in American history; it burned nearly 4 million acres, obliterated a town, and killed 1,500 people in Wisconsin in 1871. Large fires in cut-over areas and the subsequent downstream flooding were principal reasons for Congress authorizing the President in 1891 to establish forest reserves (now national forests).

Federal Fire Policy Evolution

The nascent FS focused strongly on halting wildfires in the national forests following several large fires that burned nearly 5 million acres in Montana and Idaho in 1910. The desire to control wildfires was founded on a belief that fast, aggressive control efforts were efficient, because fires that were stopped while small would not become the large, destructive conflagrations that are so expensive to control. In 1926, the agency developed its *10-acre policy*—that all wildfires should be controlled before they reached 10 acres in size—clearly aimed at keeping wildfires small. Then in 1935, the FS added its *10:00 a.m. policy*—that, for fires exceeding 10 acres, efforts should focus on control before the next burning period began (at 10:00 a.m.). These policies were seen as the most efficient and effective way to control large wildfires.[5]

In the 1970s, these aggressive FS fire control policies began to be questioned. Research had documented that, in some situations, wildfires brought ecological benefits to the burned areas— aiding regeneration of native flora, improving the habitat of native fauna, and reducing infestations of pests and of exotic and invasive species. In recognition of these benefits, the FS and the National Park Service initiated policies titled "prescribed natural fire," colloquially known as "let-burn" policies. Under these policies, fires burning within prescribed areas (such as in wilderness areas) would be monitored, rather than actively suppressed; if weather or other conditions changed or the wildfire threatened to escape the specified area, it would then be suppressed. These policies remained in effect until the 1988 wildfires in the area around Yellowstone National Park. Because at least one of the major fires in Yellowstone began as a prescribed natural fire, the agencies temporarily ended the use of the policy. Today, unplanned fire ignitions (by lightning or humans) that occur within site and weather conditions identified in fire management plans are called wildland fires for resource benefit, and are part of the agencies' fire use programs.[6]

Aggressive fire control policies were abandoned for federal wildfire planning in the late 1970s. The Office of Management and Budget challenged as excessive proposed budget increases based on these policies and a subsequent study suggested that the fire control policies would increase expenditures beyond efficient levels.[7]

Concerns about unnatural fuel loads were raised in the 1990s. Following the 1988 fires in Yellowstone, Congress established the National Commission on Wildfire Disasters, whose 1994 report described a situation of dangerously high fuel accumulations.[8] This chapter was issued shortly after a major conference examining the health of forest ecosystems in the intermountain west.[9] The summer of 1994 was another severe fire season, leading to more calls for action to prevent future severe fire seasons. The Clinton Administration developed a Western Forest Health Initiative,[10] and organized a review of federal fire policy, because of concerns that federal firefighting resources had been diverted to protecting nearby private residences and communities at a cost to federal lands and resources.[11] In December 1995, the agencies released the new *Federal Wildland Fire Management Policy & Program Review: Final Report*, which altered federal fire policy from priority for private property to equal priority for private property and federal resources, based on values at risk. (Protecting human life remains the first priority in firefighting.)

Concerns about historically unnatural fuel loads and their threat to communities persist. In 1999, the General Accounting Office (GAO; now the Government Accountability Office) issued two reports recommending a cohesive wildfire protection strategy for the FS and a combined strategy for the FS and BLM to address certain firefighting weaknesses.[12] The Clinton Administration developed a program, called the National Fire Plan, and supplemental budget request to respond to the severe 2000 fire season. In the FY200 1 Interior appropriations act (P.L. 106-291), Congress enacted the additional funding, and other requirements for the agencies.

During the severe 2002 fire season, the Bush Administration developed a proposal, called the Healthy Forests Initiative, to expedite fuel reduction projects in priority areas. The various elements of the proposal were debated, but none were enacted during the 107th Congress.[13] Some elements have been addressed through regulatory changes, while others were addressed in legislation in the 108th Congress, especially the Healthy Forests Restoration Act of 2003 (P.L. 108-148).[14]

Efficacy of Fire Protection

FS fire control programs appeared to be quite successful until the 1980s. For example, fewer than 600,000 acres of FS protected land[15] burned each year from 1935 through 1986, after averaging 1.2 million acres burned annually during the 1910s. As shown in Table 1, the average annual acreage of FS protected land burned declined nearly every decade until the 1970s, but rose substantially in the 1980s and 1990s, concurrent with the shift from fire control to fire management. Furthermore, the acreage of FS protected land burned did not exceed a million acres annually between 1920 and 1986; since then, more than a million acres of FS protected land have burned in each of at least six years—1987, 1988, 1994, 1996, 2000, and 2002. (Statistics on acreage burned by federal agency of jurisdiction have not been available from the National Interagency Fire Center since 2002.) In contrast, the acreage burned of wildlands protected by state or other federal agencies has declined substantially since the 1930s, and has continued at a relatively modest level for the past 40 years, as shown in Table 1.

There are still occasional severe fire seasons, with more than 6 million acres burned nine times since 1960 and six of those in the past decade—1963, 1969, 1996, 2000, 2002, 2004, 2005, 2006, and 2007.[16] Nonetheless, even the worst of these fire seasons (2006) saw only slightly more acres burned than the annual average in the 1950s.

Table 1. Average Annual Acreage Burned by Decade Since 1910 (in acres burned annually)

Decade	Average annual acres burned, FS Protected Lands	Average annual acres burned, Other Lands	Average annual acres burned, Total
1910-1919	1,243,572 acres	not available	not available
1920-1929	616,834 acres	25,387,733 acres	26,004,567 acres
1930-1939	343,013 acres	38,800,182 acres	39,243,195 acres
1940-1949	269,644 acres	22,650,254 acres	22,919,898 acres
1950-1959	261,264 acres	9,154,532 acres	9,415,796 acres
1960-1969	196,221 acres	4,375,034 acres	4,571,255 acres
1970-1979	242,962 acres	2,951,459 acres	3,194,421 acres
1980-1989	488,023 acres	2,494,812 acres	2,982,835 acres
1990-1999	554,577 acres	2,768,981 acres	3,323,558 acres
2000-2008	not available	not available	6 acres

Sources: U.S. Dept. of Agriculture, *Forest Service Historical Fire Statistics*, unpublished table, Washington, DC; and National Interagency Fire Center, *Fire Information—Wildland Fire Statistics*, at http://www.nifc.gov/fire fires_acres.htm, with FS acres burned deducted. (Pre-1960 data were at the NIFC site on Sept. 20, 2000, but are no longer available.)

It should also be recognized that only a small fraction of wildfires become catastrophic. In one case study, for 1986-1995 in Colorado, less than 1% of all wildfire ignitions grew to more than 1,000 acres, but these larger fires accounted for nearly 79% of the acreage burned.[17] More than 95% of the fires were less than 50 acres, and these 12,608 fires accounted for only 3% of acreage burned. Thus, a small percentage of the fires account for the vast majority of the acres burned, and probably an even larger share of the damages and control costs, since the large fires (conflagrations) burn more intensely than smaller fires and suppression costs (per acre) are higher for conflagrations because of overhead management costs and the substantial cost of aircraft used in fighting conflagrations.

Concerns and Problems

Wildfires stir a primeval fear and fascination in most of us. Many have long been concerned about the loss of valuable timber to fire and about the effects of fire on soils, watersheds, water quality, and wildlife. In addition, the loss of houses and other structures adds to wildfire damages. Historically, wildfires were considered a major threat to people and houses primarily in the brushy hillsides of southern California. However, people have increasingly been building their houses and subdivisions in forests and other wildlands, and this expanding *wildland-urban interface* has increased the wildfire threat to people and houses. Also, a century of using wildlands and suppressing wildfires has apparently significantly increased fuel loads, at least in some ecosystems, and led to historically unnatural combinations of vegetation and structures, exacerbating wildfire threats.[18]

Wildland-Urban Interface (WUI)

The wildland-urban interface has been defined as the area "where combustible homes meet combustible vegetation."[19] This interface includes a wide variety of situations, ranging from individual houses and isolated structures to subdivisions and rural communities surrounded by wildlands. While this situation has always existed to some extent, subdivisions in wildland settings appear to have grown significantly over the past two decades. Standard definitions of the interface have been developed by the federal agencies,[20] but have not been used to assess the changing situation.

Most observers agree that protecting homes and other structures in the interface is an appropriate goal for safeguarding the highest values at risk from wildfire.[21] However, there are differences of opinion about how to best protect the WUI. FS research has indicated that the characteristics of the structures and their immediate surroundings are the primary determinants of whether a structure burns. In particular, non-flammable roofs and cleared vegetation for at least 10 meters (33 feet) and up to 40 meters (130 feet) around the structure is highly likely to protect the structure from wildfire, even when neighboring structures burn.[22] Others propose reducing fuels in a band surrounding communities in the WUI; many proposals for fuel reduction suggest treatments within a half-mile (sometimes a quarter-mile) of WUI communities. Still others suggest that reducing fuels on wildlands removed from the WUI can nonetheless protect communities by reducing the danger of uncontrollable conflagrations.[23] These differences lead to discussions about the proper federal role in protecting homes in the interface (see below).

Forest and Rangeland Health

The increasing extent of wildfires in the national forests in the past two decades has been widely attributed to deteriorating forest and rangeland health, resulting at least in some cases directly from federal forest and rangeland management practices. Ecological conditions in many areas, particularly in the intermountain West (the Rocky Mountains through the Cascades and Sierra Nevadas), have been altered by various activities. Beginning more than a century ago, livestock grazing affected ecosystems by reducing the amount of grass and changing the plant species mix in forests and on rangelands. This reduced the fine fuels that carried surface fires (allowed them to spread), encouraged trees to invade traditionally open grasslands and meadows, and allowed non-native species to become established, all of which, experts believe, induce less frequent but more intense wildfires.[24] In addition, first to support mining and railroad development and later to support the wood products industry, logging of the large pines that characterized many areas has led to regeneration of smaller, less fire-resistant trees in some areas.[25] Roads that provide access for logging, grazing, and recreation have also been implicated in spreading non-native species.[26]

The nature, extent, and severity of these forest and rangeland health problems vary widely, depending on the ecosystem and the history of the site. In rangelands, the problem is likely to be invasion by non-native species (e.g., cheatgrass or spotted knapweed) or by shrubs and small trees (e.g., salt cedar or juniper). In some areas (e.g., western hemlock or inland Douglas-fir stands), the problem may be widespread dead trees due to drought or insect or disease infestations. In others (e.g., southern pines and western mixed conifers), the problem may be dense undergrowth of different plant species (e.g., palmetto in the south and firs in the west). In still others (e.g., ponderosa pine stands) the problem is more likely to be stand stagnation (e.g., too many little green trees, because intra-species competition rarely kills ponderosa pines).

One FS research report has categorized these health problems, for wildfire protection, by classifying ecosystems according to their historical fire regime.[27] The report describes five historical fire regimes:

I. ecosystems with low-severity, surface fires at least every 35 years (often called *frequent surface-fire* ecosystems);
II. ecosystems with *stand replacement* fires (killing much of the standing vegetation) at least every 35 years;
III. ecosystems with mixed severity fires (both surface and stand replacement fires) at 35- 100+ year intervals;
IV. ecosystems with stand replacement fires at 35-100+ year intervals; and
V. ecosystems with stand replacement fires at 200+ year intervals.

It is widely recognized that fire suppression has greatly exacerbated these ecological problems, at least in frequent surface-fire ecosystems (fire regime I)—forest ecosystems that evolved with frequent surface fires that burned grasses, needles, and other small fuels at least every 35 years, depending on the site and plant species (e.g., southern yellow pines and ponderosa pine). Surface fires reduce fuel loads by mineralizing biomass that may take decades to rot, and thus provide a flush of nutrients to stimulate new plant growth. Historically, many surface fires were started by lightning, although Native Americans used fires to clear grasslands of encroaching trees, stimulate seed production, and reduce undergrowth and small trees that provide habitat for undesirable insects (e.g., ticks and chiggers) and inhibit mobility and visibility when hunting.[28]

Eliminating frequent surface fires through fire suppression plus other activities has led to unnaturally high fuel loads, by historic standards, in frequent surface-fire ecosystems. These historically unnatural fuel loads can lead to stand replacement fires in ecosystems adapted to frequent surface fires. In particular, small trees and dense undergrowth can create *fuel ladders* that sometimes cause surface fires to spread upward into the forest canopy. In these ecosystems, the frequent surface fires had historically eliminated much of the understory before it got large enough to create fuel ladders. Stand replacement fires in frequent surface-fire ecosystems might regenerate new versions of the original surface-fire adapted ecosystems, but some observers are concerned that these ecosystems might be replaced with a different forest that doesn't contain the big old ponderosa pines and other traditional species of these areas.

Stand replacement fires are not, however, an ecological catastrophe in all ecosystems. Perennial grasses and some tree and brush species have evolved to regenerate following intense fires that kill much of the surface vegetation (fire regimes II, IV, and V). Aspen and some other hardwood tree and brush species, as well as most grasses, regrow from rootstocks that can survive intense wildfires. Some trees, such as jack pine in the Lake States and Canada and lodgepole pine in much of the west, have developed *serotinous* cones, that open and disperse seeds only after exposure to intense heat. In such ecosystems, stand replacement fires are normal and natural, although avoiding the incineration of structures located in those ecosystems is obviously desirable.

Some uncertainty exists over the extent of forest and rangeland health problems and how various management practices can exacerbate or alleviate the problems. In 1995, the FS estimated that 39 million acres in the National Forest System (NFS) were at high risk of catastrophic wildfire, and needed some form of fuel treatment.[29] More recently, the *Coarse-Scale Assessment* reported that 51 million NFS acres were at high risk of significant ecological damage from wildfire, and another 80 million acres were at moderate risk. (See **Table 2**.) The *Coarse-Scale Assessment* also reported 23 million acres of Department of the Interior lands at high risk and 76 million acres at moderate risk. All other lands (calculated as the total shown in the *Coarse-Scale Assessment* less the NFS and DOI lands) included 107 million acres at high risk and 314 million acres at moderate risk of ecological damage.

FUEL MANAGEMENT

Fuel management is a collection of activities intended to reduce the threat of significant damages by wildfires. The FS began its fuel management program in the 1960s. By the late 1970s, earlier agency policies of aggressive suppression of all wildfires had been modified, in recognition of the enormous cost of organizing to achieve this goal and of the ecological benefits that can result from some fires. These understandings have in particular led to an expanded prescribed burning program.

The relatively recent recognition of historically unnatural fuel loads from dead trees, dense understories of trees and other vegetation, and non-native species has spurred additional interest in fuel management activities. The presumption is that lower fuel loads and a lack of fuel ladders will reduce the extent of wildfires, the damages they cause, and the cost of controlling them. Numerous on-the-ground examples support this belief. However, little empirical research has documented this presumption. As noted in one research study, "scant information exists on fuel treatment efficacy for reducing wildfire severity."[30] This study also found that "fuel treatments moderate extreme fire behavior within treated areas, at least in" frequent surface-fire ecosystems. Others have found different results elsewhere; one study reported "no evidence that prescribed burning in these [southern California] brushlands provides any resource benefit ... in this crown- fire ecosystem."[31] A recent summary of wildfire research reported that prescribed burning generally reduced fire severity, that mechanical fuel reduction did not consistently reduce fire severity, and that little research has examined the potential impacts of mechanical fuel reduction with prescribed burning or of commercial logging.[32]

Before examining fuel management tools, a brief description of fuels may be helpful.[33] Wildfires are typically spread by fine fuels[34]—needles, leaves, grass, etc.—both on the surface and in the tree crowns (in a stand-replacement crown fire); these are known as 1-hour time lag fuels, because they dry out (lose two-thirds of their moisture content) in about an hour. Small fuels, known as 10-hour time lag fuels, are woody twigs and branches, up to a quarter-inch in diameter; these fuels also help spread wildfires because they ignite and burn quickly. Larger fuels— particularly the 1000-hour time lag fuels (more than 3 inches in diameter)—may contribute to the intensity and thus to the damage fires cause, but contribute little to the rate of spread, because they are slow to ignite. One researcher noted that only 5% of large tree stems and 10% of tree branches were consumed in high intensity fires, while 100% of the foliage and 75% of the understory vegetation were consumed.[35] Finally, *ladders* of fine and small fuels between the surface and the tree crowns can spread surface fires into the canopy, thus turning a surface fire into a stand-replacement fire.

Table 2. Lands at Risk of Ecological Change, by Historic Fire Regime (in millions of acres)

Risk of Ecological Damage	Regime I 0-35 years; surface fire	Regime II 0-35 years; crown fire	Regime III 35-100+; mixed fire	Regime IV 35-100+; crown fire	Regime V 200+ yrs; crown fire	Total
National Forest System lands						
Class 1: low	19.87	4.46	16.05	5.26	19.31	64.95
Class 2: mod.	34.96	8.66	26.71	7.35	2.76	80.45
Class 3: high	28.83	0.36	11.17	10.49	0.27	51.12
NFS Total	83.67	13.48	53.93	23.11	22.35	196.52
Department of the Interior lands						
Class 1: low	18.70	19.47	62.05	23.98	4.23	128.42
Class 2: mod.	23.83	22.87	25.82	2.93	0.38	75.83
Class 3: high	6.46	0.37	9.92	6.61	0.12	23.47
DOI Total	49.00	42.70	97.80	33.51	4.72	227.72
Private, state, and other federal lands						
Class 1: low	136.46	168.62	49.55	23.83	25.02	404.60
Class 2: mod.	117.37	101.66	59.72	25.06	10.57	313.54
Class 3: high	42.20	9.62	32.92	17.93	4.51	107.18
Other Total	296.02	279.89	142.18	66.81	40.10	825.01

Source: Schmidt et al., *Coarse-Scale Assessment,* pp. 13-15.

Prescribed Burning

Fire has been used as a tool for a long time.[36] Native Americans lit fires for various purposes, such as to reduce brush and stimulate grass growth. Settlers used fires to clear woody debris in creating agricultural fields. In forestry, fire has been used to eliminate logging debris, by burning brush piles and by prescribed burning harvested sites to prepare them for reforestation.[37]

Prescribed burning has been used increasingly over the past 40 years to reduce fuel loads on federal lands. FS prescribed burning has averaged 1.6 million acres annually over the past five years. BLM prescribed burning has averaged nearly 1.2 million acres annually since FY2003. These burning programs are a significant increase from historic levels; as recently as FY1995, the acreage in prescribed burns was 541,300 FS acres and 57,000 BLM acres. However, much of FS prescribed burning is in the FS Southern Region; prescribed burning in the intermountain west is still at relatively modest levels.

Typically, areas to be burned are identified in agency plans, and fire lines (essentially dirt paths) are created around the perimeter. The fires are lit when the weather conditions permit (i.e., when the *burning prescription* is fulfilled)—when the humidity is low enough to get the fuels to burn, but not when the humidity is so low or wind speed so high that the burning cannot be contained. (This, of course, presumes accurate knowledge of existing and expected weather and wind conditions, as well as sufficient fire control crews with adequate training on the site.) When the fire reaches the perimeter limits, the crews "mop up" the burn area to assure that no hot embers remain to start a wildfire after everyone is gone.

Prescribed burning is widely used for fuel management because it reduces biomass (the fuels) to ashes (minerals). It is particularly effective at reducing the smaller fuels, especially in the arid west where deterioration by decomposers (insects, fungi, etc.) is often very slow. In fact, it is the only human treatment that directly reduces the fine and small fuels that are important in spreading wildfires. However, prescribed fires are not particularly effective at reducing larger-diameter fuels or thinning stands to desired densities and diameters.[38]

There are several limitations in using prescribed fire. The most obvious is that prescribed fires can be risky—fire is not a *controlled* tool; rather, it is a self-sustaining chemical reaction that, once ignited, continues until the fuel supply is exhausted.[39] Fire control (for both wildfires and prescribed fires) thus focuses on removing the continuous fuel supply by creating a fire line dug down to mineral soil. The line must be wide enough to prevent the spread of fire by radiation (i.e., the heat from the flames must decline sufficiently across the space that the biomass outside the fire line does not reach combustion temperature, about 550° F). Minor variations in wind and in fuel loads adjacent to the fire line can lead to fires jumping the fire line, causing the fire to escape from control. Winds can also lift burning embers across fire lines, causing spot fires outside the fire line which can grow into major wildfires under certain conditions (such as occurred near Los Alamos, NM, in May 2000). Even when general weather conditions—temperature, humidity, and especially winds—are within the limits identified for prescribed fires, localized variations in the site (e.g., slope, aspect,[40] and fuel load) and in weather (e.g., humidity and wind) can be problematic. Thus, prescribed fires inherently carry some degree of risk, especially in ecosystems adapted to stand-replacement fires and in areas where the understory and undergrowth have created fuel ladders.

Another concern is that prescribed fires generate substantial quantities of smoke—air pollution with high concentrations of carbon monoxide, hydrocarbons, and especially particulates that degrade visibility. Some assert that prescribed fires merely shift the timing of air pollution from wildfires. Others note that smoke from pre-industrial wildland fires was at least three times more than from current levels from prescribed burning and wildfire.[41] Others have observed that fire prescriptions are typically cooler and more humid than wildfire burning conditions, and thus prescribed fires may produce more pollution (because of less efficient burning) than wildfires burning the same area. The Clean Air Act requires regulations to preserve air quality, and regulations governing particulate emissions and regional haze have been of concern to land managers who want to expand prescribed burning programs. Previous proposed legislation (e.g., H.R. 236, 106th Congress) would have exempted FS prescribed burning from air quality regulations for 10 years, to demonstrate that an aggressive prescribed burning program will reduce total particulate emissions from prescribed burning and wildfires. However, owners and operators of other particulate emitters (e.g., diesel vehicles and fossil fuel power plants) generally object to such exemptions, arguing that their emissions would likely be regulated more stringently, even though wildland fires are one of the largest sources of particulates.[42]

Salvage and Other Timber Harvesting

Another tool commonly proposed for fuel treatment is traditional timber harvesting, including salvaging dead and dying trees before they rot or succumb to disease and commercially thinning dense stands. In areas where the forest health problems include large numbers of dead and dying trees, a shift toward an inappropriate or undesirable tree species mix, or a dense understory of commercially usable trees, timber harvesting can be used to improve forest health and remove woody biomass from the forest. Nonetheless, some interest groups object to using salvage and other timber harvests to improve forest health.[43]

Timber generally may only be removed from federal forests under timber sale contracts. Stewardship contracts allow timber sales and forest management services, such as fuel reduction, to be combined in one contract, essentially as a trade of goods (timber) for services (fuel reduction); this form of contracting is discussed below, under "Other Fuel Management Tools." Because timber sale contracts have to be bought and goods-for-services contracts must generate value to provide services, the contracts generally include the removal of large, merchantable trees. Critics argue that the need for merchantable products compromises reducing fuel loads and achieving desired forest conditions.

Timber harvests remove heavy fuels that contribute to fire intensity, and can break fuel ladders, but the remaining limbs and tree tops ("slash") substantially increase fuel loads on the ground and get in the way of controlling future fires, at least in the short term, until the slash is removed or disposed of through burning. "Slash is a fire hazard mainly because it represents an unusually large volume of fuel distributed in such a way that it is a dangerous impediment in the construction of fire lines" (i.e., in suppressing fires).[44]

If logging slash is treated, as has long been a standard practice following timber harvesting, the increased fire danger from higher fuel loads that follow timber harvesting can be ameliorated. Various slash treatments are used to reduce the fire hazard, including lop-and-scatter, pile-and-burn, and chipping.[45] Lop-and-scatter consists of cutting the tops and limbs so that they lie close to the ground, thereby hastening decomposition and possibly preparing the material for broadcast burning (essentially, prescribed burning of the timber harvest site). Pile-and-burn is exactly that, piling the slash (by hand or more typically by bulldozer) and burning the piles when conditions are appropriate (dry enough, but not too dry, and with little or no wind). Chipping is feeding the slash through a chipper, a machine that reduces the slash to particles about the size of a silver dollar. and scattering the chips to allow them to decompose. Thorough slash disposal can significantly reduce fuel loads, particularly on sites with large amounts of noncommercial biomass (e.g., undergrowth and unusable tree species) and if combined with some type of prescribed burning. However, data on the actual extent of various slash disposal methods and on needed slash disposal appear to be available only for a few areas.

Other Fuel Management Tools

The other principal tool for fuel management is mechanical treatment of the fuels.[46] One common method is precommercial thinning—cutting down many of the small (less than 41/2-inch diameter) trees that have little or no current market value. Other treatments include pruning and mechanical release of seedlings (principally by cutting down or mowing competing vegetation). Mechanical treatments are often effective at eliminating fuel ladders, but as with timber cutting, do not reduce the fine fuels on the sites without additional treatment (e.g., without prescribed burning). Mechanical fuel treatments alone tend to increase fine fuels and sometimes larger fuels on the ground in the short term, until the slash has been treated.

Some critics have suggested using traditionally unused biomass, such as slash and thinning debris, in new industrial ways, such as using the wood for paper or particleboard or burning the biomass to generate electricity.[47] Research has indicated that harvesting small diameter timber may be economically feasible,[48] and one study reported net revenues of $624 per acre for comprehensive fuel reduction treatments in Montana that included removal and sale of merchantable wood.[49] However, thus far, collecting and hauling chipped slash and other biomass for products or energy have apparently not been seen as economically viable by potential timber purchasers, given that such woody materials are currently left on the harvest sites.[50]

Another possibility is to significantly change the traditional approach to timber sales. Stewardship contracting, in various forms, has been tested in various national forests.[51] Sometimes, the stewardship contract (payment and performance) is based on the condition of the stand after the treatment, rather than on the volume harvested; this is also known as end-results contracting. A variation on this theme, which has been discussed sporadically for more than 30 years, is to separate the forest treatment from the sale of the wood. The most common form is essentially to use commercial timber to pay for other treatments; that is, the contractor removes the specified commercial timber and is required to perform other activities, such as precommercial thinning of a specified area. Because of the implicit trade of timber for other activities, this is often called goods-for-services stewardship contracting. FS and BLM goods-for-services stewardship contracting was authorized through FY2013 in the FY2003 Continuing Appropriations Resolution (P.L. 108-7). Some observers believe that such alternative approaches could lead to development of an industry based on small diameter wood, and thus significantly reduce the cost of fuel management. Others fear that this could create an industry that cannot be sustained after the current excess biomass has been removed or that would need continuing subsidies.

Fuel Management Funding

Direct federal funding for prescribed burning and other fuel treatments (typically called *hazardous fuels* or *fuel management*) is part of FS and BLM appropriations for Wildfire Management. Appropriations for fuel reduction have risen from less than $100 million in FY1999 to more than $400 million annually since FY2003, and to $775 million in FY2008, with emergency supplemental funding. Funds appropriated for other purposes can also provide fuel treatment benefits. As noted above, salvage and other commercial timber sales can be used to reduce fuels in some circumstances. Various accounts, both annual appropriations and mandatory spending, provide funding for reforestation, timber stand improvement, and other activities. Reforestation actually increases fuels, but timber stand improvement includes precommercial thinning, pruning, and other mechanical vegetative treatments included in "Other Fuel Management Tools" (see above), as well as herbicide use and other treatments that do not reduce fuels.

FIRE CONTROL

Wildfire Management Funding

The cost of federal fire management is high and has risen significantly from historic levels. Wildfire appropriations for the FS and DOI totaled less than $1 billion annually prior to FY1997. For FY2003-FY2008, funding averaged more than $3 billion annually. One critic has observed that emergency supplemental appropriations, to replenish funds borrowed from other accounts to pay for firefighting, are viewed by agency employees as "free money" and has suggested that this has led to wasting federal firefighting funds, which he calls "fire boondoggles."[52] Another critic asserts that poorly designed incentives are the principal cause of the current problems and that the current fire management funding system will not resolve those problems.[53]

Over the past five years, the FS has received about 70% of the funds appropriated by Congress for wildfire preparedness and operations (including emergency supplemental funds). The other 30% goes to the BLM, which coordinates wildfire management funding for the DOI land managing agencies (BLM, the National Park Service, U.S. Fish and Wildlife Service, and Bureau of Indian Affairs); the BLM has retained about 60% of DOI funding for its wildfire activities.

Fire Control Policies and Practices

Federal fire management policy was revised in 1995, after severe fires in 1994 and the deaths of several firefighters. Current federal wildfire policy is to protect human life first, and then to protect property and natural resources from wildfires.[54] This policy includes viewing fire as a natural process in ecosystems where and when fires can be allowed to burn with reasonable safety. But when wildfires threaten life, property, and resources, the agencies act to suppress those fires.

Despite control efforts, some wildfires clearly become the kind of conflagration (stand replacement fire or crown fire) that gets media attention. As noted above, relatively few wildfires become conflagrations; it is unknown how many wildfires might become conflagrations in the absence of fire suppression efforts.

A wide array of factors determine whether a wildfire will blow up into a conflagration. Some factors are inherent in the site: slope (fires burn faster up steep slopes); aspect (south-facing slopes are warmer and drier than north-facing slopes); and ecology (some plant species are adapted to periodic stand replacement fires). Other factors are transient, changing over time (from hours to years): moisture levels (current and recent humidity; long-term drought); wind (ranging from gentle breezes to gale force winds in some thunderstorms); and fuel load and spatial distribution (more biomass and fuel ladders make conflagrations more likely).

Whether a wildfire becomes a conflagration can also be influenced by land management practices and policies. Historic grazing and logging practices (by encouraging growth of many small trees), and especially fire suppression over the past century, appear to have contributed to unprecedented fuel loads in some ecosystems. Fuel treatments can reduce fuel loads, and thus probably reduce the likelihood and severity of catastrophic wildfires, at least in some ecosystems; however, some policies and decisions may restrict fuel treatment—for example, air quality protection that limits prescribed burning or wilderness designation that prevents fuel reduction with motorized or mechanical equipment. Other practices and policies are more problematic. For example, timber harvesting can reduce fuel loads, if accompanied by effective slash disposal, but data on the need for and on the extent and efficacy of slash disposal are not available. Similarly, road construction into previously unroaded areas can increase access, and thus facilitate fuel treatment and fire suppression; conversely, roadless area protection and even road obliteration[55] can impede fuel treatment, but may reduce the likelihood of a wildfire ignition, because human-caused wildfires are more common along roads.

Once a wildfire becomes a conflagration, halting its spread is exceedingly difficult, if not impossible. Dropping water or fire retardant ("slurry") from helicopters or airplanes ("slurry bombers") can occasionally return a crown fire to the surface, where firefighters can control it, and can be used to protect individually valuable sites (e.g., structures). However, this strategy is not particularly useful in large, extended fires.[56] Setting backfires—lighting fires from a fire line to burn toward the conflagration—can eliminate the fuel ahead of the conflagration, thus halting its spread, but can be dangerous, because the backfire sometimes becomes part of the conflagration. Most firefighters recognize the futility of some firefighting efforts, acknowledging that some conflagrations will burn until they run out of fuel (move into an ecosystem or an area where the fuel is insufficient to support the conflagration) or the weather changes (the wind dies or precipitation begins, or both).

Wildfire Effects

Wildfires cause damages, killing some plants and occasionally animals.[57] Firefighters have been injured and killed, and structures can be damaged or destroyed. The loss of plants can heighten the risk of significant erosion and landslides. Some observers have reported "soil glassification," where the silica in the soils has been melted and fused, forming an impermeable layer in the soil; however, research has yet to document the extent, frequency, and duration of this condition, and the soils and burning conditions under which it occurs. Others have noted that "even the most intense forest fire will rarely have a direct heating effect on the soil at depths below 7 to 10 cm" (centimeters), about 3 to 4 inches.[58]

Damages are almost certainly greater from stand replacement fires than from surface fires. Stand replacement fires burn more fuel, and thus burn hotter (more intensely) than surface fires. Stand replacement fires kill many plants in the burned area, making natural recovery slower and increasing the potential for erosion and landslides. Also, because they burn hotter, stand replacement fires generally are more difficult to suppress, raising risks to firefighters and to structures. Finally, stand replacement fires generate substantial quantities of smoke, which can directly affect people's health and well-being.

Wildfires, especially conflagrations, can also have significant local economic effects, both short- term and long-term, with larger fires generally having greater and longer-term impacts. Wildfires, and even extreme fire danger, may directly curtail recreation and tourism in and near the fires.[59] If an area's aesthetics are impaired, local property values can decline. Extensive fire damage to trees can significantly alter the timber supply, both through a short-term glut from timber salvage and a longer-term decline while the trees regrow. Water supplies can be degraded by post-fire erosion and stream sedimentation, but the volume flowing from the burned area may increase. However, federal wildfire management includes substantial expenditures, and fire-fighting jobs are considered financially desirable in many areas.[60]

Ecological damages from fires are more difficult to determine, and may well be overstated, for two reasons. First, burned areas look devastated immediately following the fire, even when recovery is likely; for example, conifers with as much as 60% of the crown scorched are likely to survive.[61] Second, even the most intense stand replacement fires do not burn 100% of the biomass within the burn's perimeter—fires are patchy. For example, in the 1988 fires in Yellowstone, nearly 30% of the area within the fire perimeters was unburned, and another 1 5%-20% burned lightly (a surface fire); 50%-55% of the area burned as a stand replacement fire.[62]

Emergency rehabilitation is common following large fires. This is typically justified by the need for controlling erosion and preventing landslides, and may be particularly important for fire lines (dug to mineral soil) that go up steep slopes and could become gullies or ravines without treatment. Sometimes, the rehabilitation includes salvaging dead and damaged trees, because the wood's quality and value deteriorate following the fire. Emergency rehabilitation often involves seeding the sites with fast-growing grasses. While helpful for erosion control, such efforts might inhibit natural restoration if the grasses are not native species or if they inhibit tree seed germination or seedling survival.

Finally, as mentioned above, wildfires can also generate ecological benefits. Many plants regrow quickly following wildfires, because fire converts organic matter to available mineral nutrients. Some plant species, such as aspen and especially many native perennial grasses, also regrow from root systems that are rarely damaged by wildfire. Other plant species, such as lodgepole pine and jack pine, have evolved to depend on stand replacement fires for their regeneration; fire is *necessary* to open their cones and spread their seeds. One author identified research reporting various significant ecosystems threatened by *fire exclusion*—including aspen, whitebark pine, and ponderosa pine (western montane ecosystems), longleaf pine, pitch pine, and oak savannah (southern and eastern ecosystems), and the tallgrass prairie.[63] Other researchers found that, of the 146 rare, threatened, or endangered plants in the coterminous 48 states for which there is conclusive information on fire effects, 135 species (92%) benefit from fire or are found in fire- adapted ecosystems.[64]

Animals, as well as plants, can benefit from fire. Some individual animals may be killed, especially by catastrophic fires, but populations and communities are rarely threatened. Many species are attracted to burned areas following fires—some even during or immediately after the fire. Species can be attracted by the newly available minerals or the reduced vegetation allowing them to see and catch prey. Others are attracted in the weeks to months (even years) following, to the new plant growth (including fresh and available seeds and berries), for insects and other prey, or for habitat (e.g., snags for woodpeckers and other cavity nesters). A few may be highly dependent on fire; the endangered Kirtland's warbler, for example, only nests under young jack pine that was regenerated by fire, because only fire-regenerated jack pine stands are dense enough to protect the nestlings from predators.

In summary, many of the ecological benefits of wildfire that have become more widely recognized over the past 30 years are generally associated with light surface fires in frequent-fire ecosystems. This is clearly one of the justifications given for fuel treatments. Damage is likely to be greater from stand replacement fires, especially in frequent-fire ecosystems, but even crown fires produce benefits in some situations (e.g., for the jack pine regeneration needed for successful Kirtland's warbler nesting).

Roles and Responsibilities

Landowner Responsibilities

Individuals who choose to build or live in homes and other structures in the wildland-urban interface face some risk of loss from wildfires. As noted above, catastrophic fires occur, despite our best efforts, and can threaten houses and other buildings. To date, insurance companies (and state insurance regulators) have done relatively little to ameliorate these risks, in part because of federal disaster assistance paid whenever numerous homes are burned (such as in Los Alamos in May 2000). However, landowners can take steps, individually and collectively, to reduce the threat to their structures.

Research has documented that *home ignitability*—the likelihood of a house catching fire and burning down—depends substantially on the characteristics of the structure and its immediate surroundings.[65] Flammable exteriors—wood siding and especially flammable roofs—increase the chances that a structure will ignite by radiation (heat from the surrounding burning forest) or from firebrands (burning materials carried aloft by wind or convection and falling ahead of the fire). Alternate materials (e.g., brick or aluminum siding and slate or copper roofing) and protective treatments can reduce the risk. In addition, the probability of a home igniting by radiation depends on its distance from the flames. Researchers found that 85%-95% of structures with nonflammable roofs survived two major California fires (in 1961 and 1990) when there were clearances of 10 meters (33 feet) or more between the homes and surrounding vegetation.[66] Thus, building with fire resistant materials and clearing flammable materials—including vegetation, firewood piles, and untreated wood decks—from around structures reduces their chances of burning.

In addition, landowners can cooperate in protecting their homes in the wildland-urban interface. Fuel reduction within and around such subdivisions can reduce the risk, and economies of scale suggest that treatment costs for a subdivision might be lower than for an individual (especially if volunteer labor is contributed). In addition, as noted above, narrow and unmarked roads can hinder fire crews from reaching wildfires. Assuring adequate roads that are clearly marked and mapped can help firefighters to protect subdivisions. Finally, communal water sources, such as ponds and cisterns, may improve the protection of structures and subdivisions.

State and Local Government Roles and Responsibilities

In general, the states are responsible for fire protection on non-federal lands, although cooperative agreements with the federal agencies may shift those responsibilities. Typically, local governments are responsible for putting out structure fires. Maintaining some separation between suppressing structural fires and wildfires may be appropriate, because the suppression techniques and firefighter hazards and training differ substantially. Nonetheless, cooperation and some overlapping responsibilities are also warranted, simply because of the locations of federal, state, and local firefighting forces.

In addition, state and local governments have other responsibilities that affect wildfire threats to homes. For example, zoning codes—what can be built where—and building codes—permissible construction standards and materials—are typically regulated locally. These codes could (and some undoubtedly do) include restrictions, standards, or guidelines for improving fire protection in the wildland-urban interface.

The insurance industry, and home fire insurance requirements, are generally regulated by states. State regulators could work with the industry to increase the consideration of wildfire protection and home defensibility in homeowners' insurance. Road construction and road maintenance are often both state and local responsibilities, depending on the road; these roads are usually designed and identified in ways that are useful for fire suppression crews. State and local governments could further assist home protection from wildfires by supporting programs to inform residents, especially those in the urban-wildland interface, of ways that they can protect their homes.

Federal Roles and Responsibilities

The federal government has several roles in protecting lands and resources from wildfire, including protecting federal lands, assisting protection by states and local governments, and assisting public and private landowners in the aftermath of a disaster.

Federal Land Protection

The federal government clearly is responsible for fire protection on federal lands. Federal responsibility to protect neighboring non-federal lands, resources, and structures, however, is less clear. This issue was raised following several 1994 fires, where the federal officials observed that firefighting resources were diverted to protecting nearby private residences and communities at a cost to federal lands and resources.[67] In December 1995, the agencies released the new *Federal Wildland Fire Management Policy & Program Review: Final Report*, which altered federal fire policy from priority for private property to equal priority for private property and federal resources, based on values at risk. (Protecting human life is the first priority in firefighting.) Funding for fire protection of federal lands accounts for about 95% of all federal wildfire management appropriations. As noted above, fire appropriations have risen dramatically over the past decade.

Cooperative Assistance

The federal government also provides assistance for fire protection. Most federal wildfire protection assistance has been through the FS, but the Federal Emergency Management Agency (FEMA) in the Department of Homeland Security also has a program to assist in protecting communities from disasters (including wildfire).

FS efforts are operated through a cooperative fire protection program within the State and Private Forestry (S&PF) branch. This fire program includes financial and technical assistance to states and to volunteer fire departments. The funding provides a nationwide fire prevention program and equipment acquisition and transfer (the Federal Excess Personal Property program) as well as training and other help for state and local fire organizations. The 2002 Farm Bill (P.L. 107-171) created a new community fire protection program under which the FS can assist communities in fuel reduction and other activities on private lands in the wildland-urban interface. One particular program, FIREWISE, is supported through an agreement with and grant to the National Fire Protection Association, in conjunction with the National Association of State Foresters, to help private landowners learn how to protect their property from catastrophic wildfire.

Funding for cooperative fire assistance rose substantially in FY200 1, from less than $30 million to nearly $150 million. Funding has declined since, but remains substantially higher than the $15- $20 million annually in the 1990s.

FEMA has programs to assist fire protection efforts.[68] One FEMA program is fire suppression grants under the Stafford Act (the Disaster Relief and Emergency Assistance Act, 42 U.S.C. §5187). These are grants to states to assist in suppressing wildfires that threaten to become major disasters. Also, the U.S. Fire Administration is a FEMA entity charged with reducing deaths, injuries, and property losses from fires; agency programs include data collection, public education, training, and technology development.[69]

The federal government has one other program that supports federal and state wildfire protection efforts—the National Interagency Fire Center (NIFC). The center was established by the BLM and the FS in Boise, ID, in 1965 to coordinate fire protection efforts (especially aviation support) in the intermountain west. The early successes led to the inclusion of the National Weather Service (in the National Oceanic and Atmospheric Administration of the U.S. Department of Commerce) and of the other DOI agencies with fire suppression responsibilities (the National Park Service, Fish and Wildlife Service, Bureau of Indian Affairs, and Office of Aircraft Services). (FEMA is not included in the NIFC.) NIFC also coordinates with the National Association of State Foresters to assist in the efficient use of federal, state, and local firefighting resources in areas where wildfires are burning.

Disaster Relief

The federal government also provides relief following many disasters, to assist recovery by state and local governments and especially the private sector (including the insurance industry). The federal land management agencies generally do not provide disaster relief, although there has been some economic assistance for communities affected by wildfires upon occasion, as described above. Wildfire operations funding includes money for emergency rehabilitation, to reduce the possibility of significant erosion, stream sedimentation, and mass soil movement (landslides) from burned areas of federal lands. While not direct relief for affected communities, such efforts may prevent flooding and debris flows that can exacerbate local economic and social problems caused by catastrophic fires. Two authorized programs, FS Emergency Reforestation Assistance and USDA Emergency Forest Restoration, can aid private landowners whose lands were damaged by wildfire, but the programs have not been funded in recent years.[70]

FEMA is the principal federal agency that provides relief following declared disasters, although local, state, and other federal agencies (e.g., the Farm Service Agency and the Small Business Administration) also have emergency assistance programs.[71] The Stafford Act established a process for Governors to request the President to declare a disaster, and public and individual assistance programs for disaster victims.

If the risk of catastrophic fires destroying homes and communities continues to escalate, as some have suggested, requests for wildfire disaster relief would also likely rise. This might lead some to argue that a federal insurance mechanism might be a more efficient and equitable system for sharing the risk. Federal crop insurance and national flood insurance have existed for many years, while federal insurance for other catastrophic risks (e.g., hurricanes, tornados, earthquakes, volcanoes) has also been debated.[72] An analysis of these alternative systems is beyond the scope of this chapter, but these might provide alternative approaches that could be adapted for federal wildfire insurance, if such insurance were seen as appropriate. Some observers, however, object to compensating landowners for building in what critics identify as unsafe areas.[73]

CURRENT ISSUES

The severe fire seasons in recent years have raised many wildfire issues for Congress and the public. There have been spirited discussions about the effects of land management practices, especially timber sales, on fuel loads. A broad range of opinion exists on this issue, but most observers generally accept that current fuel loads reflect the aggressive fire suppression of the past century as well as historic logging and grazing practices. Some argue that catastrophic wildfires are nature's way of rejuvenating forests that have been mismanaged in extracting timber, and that the fires should be allowed to burn to restore the natural conditions.[74] Others argue that the catastrophic fires are due to increased fuel loads that have resulted from reduced logging in the national forests over the past decade, and that more logging could contribute significantly to reducing fuel loads and thus to protecting homes and communities.[75] However, the extent to which timber harvests affect the extent and severity of current and future wildfires cannot be determined from available data.[76] Some critics suggest that historic mismanagement— excessive fire suppression and past logging and grazing practices—by the FS warrants wholesale decentralization or revision of the management authority governing the National Forest System.[77]

Research information on causative factors and on the complex circumstances surrounding wildfire is limited. The value of wildfires as case studies for building predictive models is constrained, because the *a priori* situation (e.g., fuel loads and distribution) and burning conditions (e.g., wind and moisture levels, patterns, and variations) are often unknown. Experimental fires in the wild would be more useful, but are dangerous and generally unacceptable to the public. Prescribed fires could be used for research, but the burning conditions are necessarily restricted. Fires in the laboratory are feasible, but often cannot duplicate the complexity and variability of field conditions. Thus, research on fire protection and control is challenging, and predictive tools for fire protection and control are often based substantially on expert opinion and anecdotes, rather than on documented research evidence.[78]

Concerns over forest and rangeland health, particularly related to fuel loads, have been discussed for nearly two decades; a major conference on forest ecosystem health was held in Idaho in 1993.[79] Significant funding to address these concerns, however, was not proposed until September 2000. While higher funding for wildfire protection, including fuel reduction, has persisted, some question whether this additional funding is sufficient to adequately reduce fuel loads. In 1999, GAO estimated that it would cost $725 million *annually*—nearly $12 billion through 2015—to reduce fuels using traditional treatment methods on the 39 million FS acres that were estimated to be at high risk of catastrophic wildfire.[80] This is nearly double the significantly increased appropriations for FS fuel reduction since FY200 1.

The cost of a comprehensive fuel reduction program, as many advocate, would likely exceed the GAO estimate of $12 billion, because the scope of potential costs and proposed programs has increased. The FS estimate of FS acres at high risk of ecological loss due to catastrophic fire increased from 39 million acres in 1999 to 51 million acres in 2003. In addition, the GAO cost figure (received from the FS) of $300 per acre on average for fuel reduction might be low. One might anticipate more careful federal prescribed burning after the May 2000 escaped prescribed fire burned 239 homes in Los Alamos, NM; more cautious prescribed burning is likely to have higher unit costs than the GAO figure. Also, many advocate emphasizing fuel reduction in the wildland-urban interface, and treatment costs in the interface are higher, because of risks to homes and other structures from prescribed burning and because of possible damage to aesthetics from mechanical treatments.

GAO also addressed a subset of the widely-advocated comprehensive fuel reduction program, by estimating the cost for the initial treatment of FS high-risk acres. The FS has estimated that there are 23 million high-risk acres of DOI land and 107 million high-risk acres of other land. In addition, many advocate reducing fuels on lands at moderate risk—80 million FS acres, 76 million DOI acres, and 313 million other acres. Finally, in frequent-fire ecosystems, retreatment would be needed on the 5-35 year fire cycle (depending on the ecosystem), suggesting that fuel management costs would need to be continued beyond the 16-year program examined by GAO.

If a comprehensive program were undertaken to reduce fuels on all high-risk and moderate-risk federal lands, using GAO's treatment cost rate of $300 per acre, the total cost would come to $69 *billion*—$39 billion for FS lands and $30 billion for DOI lands—for initial treatment. This would come to $4.3 billion annually over 16 years, whereas the Administration's requested budget for fuel treatment in FY2008 was $499.8 million ($297.0 million for the FS and $202.8 million for the BLM), a little more than 10% of what some implicitly propose. This raises questions about whether a comprehensive fuel reduction program is feasible and how to prioritize treatment efforts.

There is a final significant question: would it work? The answer depends, in part, on how one defines successful fire protection. Fuel reduction might help restore "more natural" conditions to forests and rangelands, as many advocate, and would likely yield some social benefits (e.g., improved water quality, more habitat for fire-dependent animal species). Others, however, advocate fuel reduction to allow greater use of forests and rangelands, for timber production, recreation, water yield, etc. Fuel reduction will certainly not reduce the conflict over the goals and purposes of having and managing federal lands. Reducing fuel loads might reduce acreage burned and the severity and damages of the wildfires that occur. Research is needed in various ecosystems to document and quantify the relationships among fuel loads and damages and the probability of catastrophic wildfires, to examine whether the cost of fuel reduction is justified by the lower fire risk and damage. However, it should also be recognized that, regardless of the extent of fuel reduction and other fire protection efforts, as long as there is biomass for burning, especially under severe weather conditions (drought and high wind), catastrophic wildfires will occasionally occur, with the attendant damages to resources, destruction of nearby homes, other economic and social impacts, and potential loss of life.

REFERENCES

Agee, James K. *Fire Ecology of Pacific Northwest Forests*. Washington, DC: Island Press, (1993). 493 p.

Brown, Arthur A. & Davis, Kenneth P. (1973). *Forest Fire Control and Use*. 2nd ed. New York, NY: McGraw-Hill Book company,. 686 p.

Carle, David. (2002). *Burning Questions: America's Fight With Nature's Fire*. Westport, CT: Praeger Publishers,. 298 p.

Chandler Craig, Cheney Phillip, Thomas Philip, Trabaud Louis & Williams Dave. (1983). *Fire In Forestry. Volume I: Forest Fire Behavior and Effects*. New York, NY: John Wiley & Sons. 450 p.

Chandler Craig, Cheney Phillip, Thomas Philip, Trabaud Louis & Williams Dave. (1983). *Fire In Forestry. Volume II: Forest Fire Management and Organization*. New York, NY: John Wiley & Sons, 298 p.

Gonzalez-Caban, Armando & Omi, Philip N. (1999). technical coordinators. *Proceedings of the Symposium on Fire Economics, Planning, and Policy: Bottom Lines*. General Technical Report PSW-GTR-173. Berkeley, CA: USDA Forest Service, Pacific Southwest Research Station, Dec. 332 p.

Kozlowski, T. T. & Ahlgren, C. E. (1974). eds. *Fire and Ecosystems*. New York, NY: Academic Press,. 542 p.

National Academy of Public Administration. (2002). *Wildfire Suppression: Strategies for Containing Costs*. Washington, DC: Sept.. 2 volumes.

Nelson, Robert H. (2000). *A Burning Issue: A Case for Abolishing the U.S. Forest Service*. Lanham, MD: Rowman & Littlefield Publishers, Inc.,. 191 p.

O'Toole, Randal. *Reforming the Fire Service: An Analysis of Federal Fire Budgets and Incentives*. Bandon, OR: Thoreau Institute, July 2002. 53 p.

Pyne, Stephen J. (1982). *Fire in America: A Cultural History of Wildland and Rural Fire*. Princeton, NJ: Princeton University Press, 654 p.

Pyne, Stephen J., Andrews, Patricia L. & Laven, Richard D. (1996). *Introduction to Wildland Fire*, 2nd ed. New York, NY: John Wiley & Sons, Inc.,. 769 p.

Sampson, R. Neil & Adams David L. (1994). eds. *Assessing Forest Ecosystem Health in the Inland West: Papers from the American Forests Workshop, November 14th-20th, 1993, Sun Valley, Idaho*. New York, NY: Food Products Press,. 461 p.

Wright, Henry A. & Bailey, Arthur W. *Fire Ecology: United States and Southern Canada.* New York, NY: John Wiley & Sons, 1982. 501 p.

End Notes

[1] Wildlands is a term commonly used for undeveloped areas—forests, grasslands, brush fields, wetlands, deserts, etc. It excludes agricultural lands and pastures, residential areas, and other, relatively intensively developed areas.

[2] National Interagency Fire Center, "Fire Information—Wildland Fire Statistics," available at http://www.nifc.gov/fire_info/fires_acres.htm. Fire season severity is commonly assessed by acres burned, but larger fires may not be "worse" if they burn less intensely, because their damages may be lower. However, fire intensity and damages are not measured for each wildfire, and thus cannot be used to gauge the severity of fire seasons. It is uncertain whether acreage burned might be a reasonable approximation of severity.

[3] In 2003, there were 810.7 million acres of *private* forests and rangelands in the coterminous 48 states. (U.S. Dept. of Agriculture, Natural Resources Conservation Service, *National Resources Inventory: 2003 Annual NRI*, February 2007, p. 5.) This is substantially more than the 386.1 million acres of lands in those 48 states administered by the FS and DOI. (See CRS Report RL32393, *Federal Land Management Agencies: Background on Land and Resources Management*, by Carol Hardy Vincent et al.)

[4] See also CRS Report RS21880, *Wildfire Protection in the Wildland-Urban Interface*, by Ross W. Gorte.

[5] See Julie K. Gorte and Ross W. Gorte, *Application of Economic Techniques to Fire Management—A Status Review and Evaluation*, USDA Forest Service, Intermountain Research Station, Gen. Tech. Rept. INT-53, Ogden, UT, June 1979.

[6] U.S. Dept. of the Interior and Dept. of Agriculture, *Federal Wildland Fire Management Policy & Program Review: Final Report*, Washington, DC, Dec. 18, 1995. Hereafter referred to as *1995 Federal Wildland Fire Review*.

[7] Stephen J. Pyne, *Fire In America: A Cultural History of Wildland and Rural Fire*, Princeton Univ. Press, Princeton, NJ, 1982, pp. 293-294.

[8] R. Neil Sampson, chair, *Report of the National Commission on Wildfire Disasters*, Washington, DC, 1994.

[9] See R. Neil Sampson and David L. Adams, eds., *Assessing Forest Ecosystem Health in the Inland West: Papers from the American Forests Workshop, November 14th-20th, 1993, Sun Valley, Idaho*, Food Products Press, New York, NY, 1994. Hereafter cited as *Assessing Forest Ecosystem Health in the Inland West.*

[10] U.S. Dept. of Agriculture, Forest Service, State and Private Forestry, *Western Forest Health Initiative*, Washington, DC, October 31, 1994.

[11] Bob Armstrong, Assistant Secretary for Lands and Minerals Management, U.S. Dept. of the Interior, "Statement," *Fire Policy and Related Forest Health Issues*, joint oversight hearing, House Committees on Resources and on Agriculture, October 4, 1994, U.S. GPO, Washington, DC, 1995, p. 9. Serials No. 103-119 (Committee on Resources) and 103-82 (Committee on Agriculture).

[12] U.S. General Accounting Office, *Western National Forests: A Cohesive Strategy is Needed to Address Catastrophic Wildfire Threats*, GAO/RCED-99-65, Washington, DC, April 1999; and *Federal Wildfire Activities: Current Strategy and Issues Needing Attention*, GAO/RCED-99-233, Washington, DC, August 1999. Hereafter cited as GAO, *Cohesive Strategy Needed.*

[13] See out-of-print CRS Report RL3 1679, *Wildfire Protection: Legislation in the 107th Congress and Issues in the 108th Congress*, by Ross W. Gorte (available from the author).

[14] For information on recent regulatory and legislative developments on wildfire protection, see CRS Report RL33792, *Federal Lands Managed by the Bureau of Land Management (BLM) and the Forest Service (FS): Issues for the 110th Congress*, by Ross W. Gorte et al.

[15] Under several cooperative agreements, developed to improve protection efficiency, the Forest Service protects some non-federal lands, while other organizations protect some national forest lands; the total acres protected by the Forest Service roughly equals the acres in the National Forest System.

[16] The National Interagency Fire Center has revised the data for 1983-2002, dropping 1988 (the year of the Yellowstone fires) off the list.

[17] Leon F. Neuenschwander et al., "Indexing Colorado Watersheds to Risk of Wildfire," *Mapping Wildfire Hazards and Risks*, Food Products Press, New York, NY,2000, pp. 35-55.

[18] R. Neil Sampson et al., "Assessing Forest Ecosystem Health in the Inland West: Overview," *Assessing Forest Ecosystem Health in the Inland West*, pp. 3-10.

[19] *Wildfire Strikes Home! The Report of the National Wildland/Urban Fire Protection Conference*, sponsored by the USDA, Forest Service; the National Fire Protection Association; and the FEMA, U.S. Fire Administration, January 1987, p. 2.

[20] U.S. Dept. of Agriculture and Dept. of the Interior, "Urban Wildland Interface Communities Within the Vicinity of Federal Lands That Are at High Risk From Wildfire," 66 *Federal Register* 751-754, January 4, 2001.

[21] See CRS Report RL34517, *Wildfire Damages to Homes and Resources: Understanding Causes and Reducing Losses*, by Ross W. Gorte

[22] Jack D. Cohen, "Preventing Disaster: Home Ignitability in the Wildland-Urban Interface," *Journal of Forestry*, vol. 102, no. 3 (March 2000), pp. 15-21.

[23] Personal communication, Harv Forsgren, Regional Forester (Region 3), USDA Forest Service in Washington, DC, on Aug. 21, 2003.

[24] W. W. Covington and M. M. Moore, "Postsettlement Changes in Natural Fire Regimes and Forest Structure: Ecological Restoration of Old-Growth Ponderosa Pine Forests," *Assessing Forest Ecosystem Health in the Inland West*, pp. 153-181.

[25] Jay O'Laughlin, "Assessing Forest Health Conditions in Idaho with Forest Inventory Data," *Assessing Forest Ecosystem Health in the Inland West*, pp. 221-247.

[26] Federal Interagency Committee for the Management of Noxious and Exotic Weeds, *Invasive Plants: Changing the Landscape of America*, Washington, DC, 1998, pp. 23-24.

[27] Kirsten M. Schmidt et al., *Development of Coarse-Scale Spatial Data for Wildland Fire and Fuel Management*, USDA Forest Service, Rocky Mountain Research Station, Gen. Tech. Rept. RMRS-87, Ft. Collins, CO, April 2002. Hereafter cited as Schmidt et al., *Coarse-Scale Assessment.*

[28] James K. Agee, *Fire Ecology of Pacific Northwest Forests*, Island Press, Washington, DC, 1993, pp. 54-57. Hereafter cited as Agee, *Fire Ecology of PNW Forests.*

[29] Enoch Bell et al., *Fire Economics Assessment Report*, unpublished report submitted to USDA Forest Service, Fire and Aviation Management, on Sept. 1, 1995.

[30] Philip N. Omi and Erik J. Martinson, *Effects of Fuels Treatment on Wildfire Severity: Final Report*, submitted to the Joint Fire Science Program Governing Board, Colorado State Univ., Western Forest Fire Research Center, Ft. Collins, CO, March 25, 2002, p. i.

[31] Jon E. Keeley, "Fire Management of California Shrubland Landscapes," *Environmental Management*, vol. 29, no. 3 (2002), pp. 3 95-408.

[32] Henry Carey and Martha Schumann, *Modifying WildFire Behavior—The Effectiveness of Fuel Treatments: The Status of Out Knowledge*, Southwest Region Working Paper 2, National Community Forestry Center, Santa Fe, NM, April 2003.

[33] See Arthur A. Brown and Kenneth P. Davis, "Chapter 4: Forest Fuels," *Forest Fire Control and Use*, McGraw-Hill Book Co., New York, NY, 1973, pp. 79-110. Hereafter cited as Brown and Davis, *Fire Control and Use.*

[34] Robert E. Martin and Arthur P. Brackebusch, "Fire Hazard and Conflagration Prevention," *Environmental Effects of Forest Residues Management in the Pacific Northwest: A State-of-Knowledge Compendium,* USDA Forest Service, Pacific Northwest Research Station, Gen. Tech. Rept. PNW-24, Portland, OR, 1974.

[35] Agee, *Fire Ecology of PNW Forests*, p. 42. It is also important to recognize that the percentage of biomass in 1-hour, 10-hour, 100-hour, and 1000-hour fuels depends largely on tree diameter, with the percentage in large fuels increasing as diameter increases.

[36] Historical evidence indicates that current levels of burning through prescribed burns and wildfires represent levels perhaps 10%-30% of pre-industrial burning levels from natural and Native-set fires. See Bill Leenhouts, "Assessment of Biomass Burning in the Conterminous United States," *Conservation Ecology* 2(1), 1998, available on Jan. 16, 2007, at http://www.ecologyandsociety.org/vol2/iss1/art1/. Hereafter cited as Leenhouts, *Assessment of Biomass Burning.*

[37] David M. Smith et al., *The Practice of Silviculture: Applied Forest Ecology*, John Wiley & Sons, New York, NY, 1997. Hereafter cited as Smith et al., *The Practice of Silviculture.*

[38] See Brown and Davis, *Fire Control and Use*, pp. 560-572.

[39] Fire can also be halted by eliminating the supply of oxygen, as occurs when fire retardant ("slurry") is spread on forest fires from airplanes ("slurry bombers"). However, reducing oxygen supply usually can only occur in a limited area, because of the cost to spread the fire retardant.

[40] Aspect is the direction which the slope is facing; in the northern hemisphere, south-facing slopes (south aspects) get more radiant energy from the sun than north aspects, and thus are inherently warmer and drier, and hence are at greater risk of more intense wildfires.

[41] Leenhouts, *Assessment of Biomass Burning.*

[42] See, for example, U.S. House, Committee on Resources, *Hearing on the Use of Fire as a Management Tool and Its Risks and Benefits for Forest Health and Air Quality*, Sept. 30, 1997, Serial No. 105-45, GPO, Washington, DC, 141 p.

[43] Timber harvesting has a variety of proponents and opponents for reasons beyond fuel management. Some interests object to timber harvesting on a variety of grounds, including the poor financial performance of FS timber sales and the degradation of water quality and certain wildlife habitats that follows some timber harvesting. Others defend timber sales for the employment and income provided in isolated, resource-dependent communities as well as for increasing water yields and available habitat for other wildlife species. The arguments supporting and opposing timber harvests generally have often been raised in discussions about

fire protection, but are not reproduced in this report. See CRS Report 95-364, *Salvage Timber Sales and Forest Health*, by Ross W. Gorte (out-of-print; available from the author).

[44] Smith et al., *The Practice of Silviculture*.

[45] Ibid.

[46] Chemical treatments (herbicides) are also used in forestry, mostly on unwanted vegetation, but they are not included here as a fuel treatment tool, because they are used primarily to kill live biomass rather than to reduce biomass levels on a site. Biological treatments (e.g., using goats to eat the small diameter material) are feasible, but are rarely used.

[47] Robert Nelson, University of Maryland, cited in: Rocky Barker, "Wildfires Creating Odd Bedfellows," *The Idaho Statesman*, Aug. 14, 2000, pp. 1A, 7A.

[48] Henry Spelter, Ron Wang, and Peter Ince, *Economic Feasibility of Products From Inland West Small Diameter Timber*, USDA Forest Service, Forest Products Lab, FPL-GTR-92, Madison, WI, May 1996, 17 p.

[49] Carl E. Fieldler, Charles E. Keegan, Todd A. Morgan, and Christopher W. Woodall, "Fire Hazard and Potential Treatment Effectiveness: A Statewide Assessment in Montana," *Journal of Forestry*, vol. 101, no. 2 (March 2003), p. 7.

[50] Research documenting the economics of slash use (in contrast to small diameter trees) is lacking. However, this seems a reasonable conclusion, given that the slash is left on the site by the timber purchaser (who could remove and sell the material) and that the agencies and various interest groups have been trying to develop alternatives to the traditional contracts (e.g., stewardship contracts) to remove thinning slash and other biomass fuels.

[51] See CRS Report RS20985, *Stewardship Contracting for Federal Forests*, by Ross W. Gorte.

[52] Robert H. Nelson, *A Burning Issue: A Case for Abolishing the U.S. Forest Service*, Rowman & Littlefield Publishers, Inc., Lanham, MD, 2000, pp. 15-43. Hereafter cited as Nelson, *A Burning Issue*.

[53] Randal O'Toole, *Reforming the Fire Service: An Analysis of Federal Fire Budgets and Incentives*, Thoreau Institute, Bandon, OR, July 2002. Hereafter cited as O'Toole, *Reforming the Fire Service*.

[54] *1995 Federal Wildland Fire Review*.

[55] Road obliteration is closing the road and returning the roadbed to near-natural conditions.

[56] *Federal Aerial Firefighting: Assessing Safety and Effectiveness*, Blue Ribbon Panel Report to the Chief, USDA Forest Service and Director, USDI Bureau of Land Management, available at http://www.wildfirelessons.net/documents/BRP_Final12052002.pdf.

[57] For a thorough discussion of these effects, see L. Jack Lyon et al., *Wildland Fire in Ecosystems: Effects of Fire on Fauna*, USDA Forest Service, Rocky Mountain Research Station, Gen. Tech. Rept. RMRS-GTR-42-vol. 1 ,Ogden, UT, Jan. 2000. Hereafter cited as Lyon, et al., *Effects of Fire on Fauna*.

[58] Craig Chandler et al., *Fire In Forestry. Volume I: Forest Fire Behavior and Effects*, John Wiley & Sons, New York, NY, 1983, p. 173.

[59] Paul E. Polzin, Michael S. Yuan, and Ervin G. Schuster, *Some Economic Impacts of the 1988 Fires in the Yellowstone Area*, USDA Forest Service, Intermountain Research Station, Research Note INT-41 8, Ogden, UT, October 1993.

[60] Nelson, *A Burning Issue*, pp. 37-38.

[61] See Ross W. Gorte, *Fire Effects Appraisal: The Wisconsin DNR Example*, Michigan State Univ., Ph.D. dissertation, East Lansing, MI, June 1981.

[62] See Lyon, et al., *Effects of Fire on Fauna*, p. 44.

[63] Leenhouts, *Assessment of Biomass Burning*.

[64] Amy Hessl and Susan Spackman, *Effects of Fire on Threatened and Endangered Plants: An Annotated Bibliography*, U.S. Dept. of the Interior, National Biological Service, Information and Technical Report 2, Fort Collins, CO, n.d.

[65] See Jack D. Cohen, "Reducing the Wildland Fire Threat to Homes: Where and How Much?" *Proceedings of the Symposium on Fire Economics, Planning, and Policy: Bottom Lines* (San Diego, CA: April 5-9, 1999), USDA Forest Service, Pacific Southwest Research Station, Gen. Tech. Rept. PSW-GTR-173, Berkeley, CA, Dec. 1999, pp. 189-195. Hereafter cited as Cohen, *Reducing the Wildland Fire Threat to Homes*.

[66] Ibid.

[67] See footnote 11.

[68] The annual funding for these programs is not distinguished in the agency's annual budget justification, and thus is not included in this report. See CRS Report RL34004, *Homeland Security Department: FY2008 Appropriations*, coordinated by Jennifer E. Lake.

[69] See CRS Report RS20071, *United States Fire Administration: An Overview*, by Lennard G. Kruger.

[70] See CRS Report RL3 1065, *Forestry Assistance Programs*, by Ross W. Gorte, p. 10.

[71] See CRS Report RL3 1734, *Federal Disaster Recovery Programs: Brief Summaries*, by Mary B. Jordan.

[72] See CRS Report RS21212, *Agricultural Disaster Assistance*, by Ralph M. Chite, and CRS Report RL34367, *Side-bySide Comparison of Flood Insurance Reform Legislation in the 110th Congress*, by Rawle O. King

[73] Personal communication with Tim Hermach, Founder and President, Native Forest Council, Eugene, OR, on Oct. 18, 2000.

[74] Personal communication with Tim Hermach, Founder and President, Native Forest Council, Eugene, OR, on Sept. 26, 2000.

[75] William N. Dennison, Plumas County Supervisor, District 3, "Statement," *Hearing on the Use of Fire as a Management Tool and Its Risks and Benefits for Forest Health and Air Quality*, House Committee on Resources, Sept. 30, 1997, Serial No. 105-45, GPO, Washington, DC, 1997, pp. 107-116.

[76] See CRS Congressional Distribution Memorandum, *Forest Fires and Forest Management*, by Ross W. Gorte, Sept. 20, 2000.

[77] Nelson, *A Burning Issue*; O'Toole, *Reforming the Fire Service*.

[78] Fire experts typically believe (and must believe, to do their jobs effectively) that catastrophic wildfires can and should be controlled; thus, their opinions may be biased, overstating the effectiveness and efficiency of control efforts.

[79] *Assessing Forest Ecosystem Health in the Inland West: November 14th-20th, 1993*. See footnote 8.

[80] GAO, *Cohesive Strategy Needed.*

In: Wildfires and Wildfire Management
Editor: Kian V. Medina

ISBN: 978-1-60876-009-1

Chapter 4

WILDFIRE DAMAGES TO HOMES AND RESOURCES: UNDERSTANDING CAUSES AND REDUCING LOSSES

Ross W. Gorte

SUMMARY

Wildfires are getting more severe, with more acres and houses burned and more people at risk. This results from excess biomass in the forests, due to past logging and grazing and a century of fire suppression, combined with an expanding wildlandurban interface — more people and houses in and near the forests — along with climate change, exacerbating drought and insect and disease problems. Some assert that current efforts to reduce biomass (fuel treatments, such as thinning) and to protect houses are inadequate, and that public objections to activities on federal lands raise costs and delay action. Others counter that proposals for federal lands allow timber harvesting, with substantial environmental damage and little fire protection. Congress is addressing these issues through various legislative proposals and through funding for protection programs.

Wildfires are inevitable — biomass, dry conditions, and lightning create fires. Some are *surface fires*, burning needles, grasses, and other fine fuels and leaving most trees alive. Others are *crown fires*, burning biomass at all levels from the ground through the tree tops, and typically driven by high winds. Many wildfires contain areas of both surface and crown fires. Surface fires are relatively easy to control, but crown fires are difficult, if not impossible, to stop; often, crown fires burn until they run out of fuel or the weather changes.

Homes can be ignited by direct contact with the fire, by radiative heating, and by *firebrands* (burning materials lifted by the wind or the fire's own convection column). Protection of homes must address all three. Research has identified the keys to protecting structures: having a non-flammable roof; clearing burnable materials that abut the house (e.g., plants, flammable mulch, woodpiles, wooden decks); and landscaping to create a defensible space around the structure.

Wildland and resource damages from fire vary widely, depending on the nature of the ecosystem as well as on site-specific conditions. Surface-fire ecosystems, which burned on 5-

to 35-year cycles, can be damaged by crown fires, due to unnatural fuel accumulations and *fuel ladders*; fuel treatments probably prevent some crown fires in such ecosystems. Stand-replacement-fire ecosystems are where crown fires are natural and the species are adapted to periodic crown fires; fuel treatments are unlikely to alter the historic fire regime of such ecosystems. In mixed-intensityfire ecosystems, where a mix of surface and crown fires is historically normal, it is unclear whether fuel treatments would alter wildfire patterns.

Prescribed burning (intentional fires) and mechanical treatments (cutting and removing some trees) can reduce resource damages caused by wildfires in some ecosystems. However, prescribed fires are risky, mechanical treatments can cause other ecological damages, and both are expensive. Proponents of more treatment advocate expedited processes for environmental and public review of projects to hasten action and cut costs, but others caution that inadequate review can allow unintended damages with few fire protection benefits.

Wildfires have been getting more severe in recent fire seasons; the last four seasons are the most severe since 1960.[1] National attention was focused on the growing wildfire problem by an escaped prescribed fire that burned 239 houses in Los Alamos, NM, in May 2000. The fire in Los Alamos highlighted the *wildlandurban interface* problem. Also, at that time, 2000 was the second most severe fire season since 1960, eclipsed only by the 1988 Yellowstone fires.[2] President Clinton responded with a new National Fire Plan to increase funding for wildfire protection.

It has been widely proclaimed that the increasing severity of wildfires is a result of excessive biomass accumulations. In at least some ecosystems, logging, livestock grazing, and a century of fire suppression efforts have allowed biomass fuels to accumulate to unnatural levels. Climate change, and its impacts on drought, fire, and insects and diseases, could exacerbate these problems. Many interests have proposed fuel reduction treatments as a means to lower the fuel levels and thus reduce the wildfire threat to homes and to wildlands. The severe 2002 fire season led President Bush to propose a Healthy Forests Initiative to expedite efforts to reduce biomass fuels on federal lands, and in 2003, Congress enacted the Healthy Forests Restoration Act to expedite federal fuel reduction and other forest protection programs.

Some interests are concerned that current efforts to reduce fuel levels on federal lands are inadequate, and that "environmentalist objections" to some of those efforts are unnecessarily raising costs and delaying action. Others counter that some efforts are so broad that they permit substantial timber sales without significantly reducing wildfire risks for communities. Congress continues to address these issues as it considers funding and legislative proposals.

This chapter focuses on options for protecting structures and for protecting wildlands and natural resources from wildfires. It begins with a brief overview of the nature of wildfires, followed by a discussion of protecting structures. Then, it discusses wildfire damages to wildlands and natural resources, fuel treatment options and their benefits and limitations, and public involvement in federal decisions.

BACKGROUND: FIRES HAPPEN

In temperate ecosystems, wildfires are inevitable. The combination of biomass plus dry conditions — in the short term (e.g., the annual dry season) or in the long term (e.g., drought or climate change) — equals fuel to burn. Add an ignition source, such as lightning, and wildfire happens. Fire is a self-sustaining chemical reaction that perpetuates itself as long as all three elements of the fire triangle — fuel, heat, and oxygen — remain available. Fire control focuses on removing one of those elements.

There are two principal kinds of wildfire, although an individual wildfire may contain areas of both kinds.[3] One is a *surface fire*, which burns the needles or leaves, grass, and other small biomass within a foot or so of the ground and quickly moves on. Such fires are relatively easy to control by removing fuel with a *fireline,* essentially a dirt path wide enough to eliminate the continuous fuels needed to sustain the fire, or by cooling or smothering the flames with water or dirt.

The other principal kind of wildfire is a *crown fire*, also called a conflagration. Crown fires burn biomass at all levels — from the surface through the tops of the crowns of the trees — although they do not consume all the biomass; logs and large limbs may need to burn for hours before being completely reduced to ashes. Rather, a crown fire quickly burns the needles or leaves and small twigs and limbs on the surface and throughout the crown of the trees. Because the needles and leaves in the crown are green, they require more energy to burn than dry fuels on the surface. Furthermore, because of the green fuels and the often discontinuous biomass of the canopy, wind is usually needed to sustain a crown fire. Once burning vigorously, a crown fire can create its own wind (the strong upward convection of the heated air can draw in cooler air from surrounding areas, thus creating a wind that feeds the fire). The strong upward convection can also lift burning biomass (*firebrands*) and send it soaring ahead of the fire, creating spot fires and accelerating the spread of the wildfire.

Not surprisingly, crown fires are difficult, if not impossible, to control. Unless quite wide, firelines may be ineffective, especially if winds are causing spot fires. Water or fire retardant (*slurry*) dropped from helicopters or airplanes can sometimes knock a crown fire down (back to a surface fire) if the area burning and the winds are not too great. Oftentimes, however, crown fires burn until they run out of fuel or the weather changes (the wind dies or it rains or snows).

Crown fires also typically include areas of surface fire and unburned areas within their perimeters. Nearly all fires are "patchy," with a mix of areas of varying fire severities, depending on site-specific fuel, moisture, and wind conditions. This patchiness makes understanding and controlling wildfires difficult, at best.

PROTECTING STRUCTURES FROM WILDFIRES

Wildfires occasionally burn houses, in a zone commonly called the *wildlandurban interface.*[4] In recent years, it seems one or more fires annually have burned down several to few hundred homes and outbuildings (sheds, garages, etc.). These structures generally have ignited in one of three ways: through direct contact with fire, through radiation (heating from exposure to flames), and through firebrands.[5] The likelihood of a structure burning from one of these ignition methods is called *home ignitability.*[6]

Home Ignitability

Research has identified three essential elements to protecting structures: the roof; adjacent burnable materials; and the landscaping. Treating these three elements address all three ways by which structures are ignited — direct contact, radiation, and firebrands.

The roof is critical to protecting structures from wildfires.[7] Firebrands that land on a flammable roof can ignite the roof. Untreated red cedar shakes and shingles are particularly problematic: "A major cause of home loss in wildland areas is flammable woodshake roofs."[8] Fire retardant treatments offer sufficient for wood shakes, but the effectiveness of such treatments degrades over time.[9] Alternatives include tile, slate, metals (e.g., copper or aluminum), and other non-flammable materials. Walls, doors and windows, and vents can also contribute to the protection, or destruction, of a structure, depending on materials, location, and other variables.[10]

Adjacent burnable materials are items which can burn that abut the house. This can include plants (live or dead) and flammable mulch (e.g., wood chips or bark) under an overhang or eave or next to the structure, gutters clogged with leaves or needles, decks and porches, sheds and garages, and especially woodpiles. These factors were particularly important for the 239 homes burned by the Cerro Grande fire in Las Alamos, NM, in May 2000: "... the high ignitability of Los Alamos was principally due to the abundance and ubiquity of pine needles, dead leaves, cured vegetation, flammable shrubs, and wood piles that were adjacent to, touching, or covering parts of homes."[11] One source recommended that "when assessing the ignition potential of a structure, attachments [such as decks, porches, and fences] are considered part of the structure."[12]

Finally, landscaping — the characteristics of the vegetation surrounding the house — is critical to preventing both direct burning and ignition by radiation. Recommended *defensible space* around structures is at least 30 feet or 10 meters, with greater distances for steeper slopes (because of up-slope convection heating) and for larger vegetation (least for grass, more for shrubs, most for mature forest).[13] Others recommend greater distances, such as 100 feet.[14] One researcher calculated the ignition time for an untreated wood wall was more than 10 minutes at a distance of 40 meters (about 130 feet).[15] With burning durations for crown fires "on the order of 1 minute at a specific location," the "safe distance" for an untreated wood wall was calculated to be 27 meters (less than 90 feet), which is consistent with field tests documenting wall ignition times for experimental crown fires in Canada.[16] The same source notes older fire case studies documenting structure survival — 95% survival for 10-18 meter (about 32-60 feet) clearance in the 1961 Belair-Brentwood (CA) Fire and 86% survival for 10+ meter clearance in the 1990 Painted Cave (CA) Fire.[17] Thus, clearing to 40 meters would likely be considered ideal, to 30 meters desirable, and to at least 10 meters essential to achieving about 90% probability of survival. Note also that "clearing a defensible space" does not require an expanse of concrete or gravel; relatively non-flammable vegetation, such as a lawn or succulent, herbaceous plants and flowers, can provide comparable protection.[18]

The importance of landscapes in protecting structures can also be deduced from evidence from the 2002 Hayman Fire, the largest wildfire in Colorado history. A total of 132 homes were burned in the Hayman Fire. Of these, 70 (53%) "were destroyed in association with the occurrence of torching or crown fire in the home ignition zone. Sixty-two [47%] were destroyed by surface fire or firebrands."[19] Conversely, 662 homes — 83% of the homes within the fire perimeter — survived the Hayman Fire relatively unscathed.[20] Since 35% of the Hayman Fire was a high- severity burn, and another 16% was a moderate-severity burn,[21] it seems likely that at least some of these homes (the number and portion are not documented) survived despite crown fire around them. Thus, it seems reasonable to conclude that the nature of the structure — rather the nature of the fire — primarily determines whether a structure survives a wildfire.

Responsibility for Protecting Structures

Owners are responsible for their structures. Insurance companies and the relevant state agencies that regulate insurance can contribute to structural protection by requiring certain materials and actions to obtain a policy for compensation following wildfire losses or by adjusting premiums based on homeowner actions. Local governmental agencies also play a role, since the building and zoning codes that could implement some of the safe- structure requirements are generally developed and enforced locally.[22] Alternatively, states can play a role; as of January 1, 2008, the California Building Standards Commission[23] is enforcing wildland-urban interface building standards in very high hazard zones.

The structure owners are also primarily responsible for the defensible space surrounding their structures. A 10-meter-wide clearing around a 3,000 square-foot structure encompasses less than a third of an acre — almost certainly private land owned in conjunction with the structure. Even a 40-meter clearing encompasses less than 2 acres, and thus is commonly part of the structure owner's property in the wildland-urban interface.

When a structural fire starts, the local fire department is responsible for controlling the blaze. State agencies may provide support for local fire departments, especially in the wildland-urban interface where a structural fire could cause a wildland fire. Occasionally, because of the location of fire-fighting resources, the federal agencies may be the first responders on a structural fire in the interface, but federal fire-fighters are generally not trained for safety in structural fire-fighting. The federal government has no responsibility for structural fire control in the wildland- urban interface. However, the Forest Service (FS) does have programs to provide technical and financial assistance to states and to volunteer fire departments.[24]

Given the nature of efforts needed to protect structures and that developing, adopting, and enforcing building codes are local and state responsibilities, there is no clear federal responsibility in protecting structures from wildfires. However, the federal government often provides disaster assistance in the wake of a catastrophic wildfire, generally at the request of a governor. Federal disaster assistance is expensive and could be avoided if action to protect homes were taken in advance. Several federal agencies currently support FIREWISE, a program aimed at educating homeowners about how to make their structures safe from wildfire.[25] Assistance to homeowners — such as technical assistance, low-cost loans, and cost-sharing on projects — might be a cost-saving federal investment. Federal assistance to prepare local firefighters is another means for addressing home protection from wildfires.[26] Research on wildland-urban interface fire protection can also reduce losses.[27] Another possibility might be federal wildfire insurance, comparable to the National Flood Insurance Program.[28] Those living in an identified wildfire-prone zone would be required to purchase federal wildfire insurance (probably with an annual premium) to receive compensation for wildfire damages. The premiums could vary by ecoregion, depending on the likelihood and risk of wildfires, and by aspects of the structure and landscaping (which might require periodic inspections).

PROTECTING WILDLANDS AND NATURAL RESOURCES

Wildlands and natural resources can also be damaged by wildfires. Wildfire damages vary widely, depending on the nature of the ecosystems burned as well as site-specific conditions. Activities to modify wildland biomass fuels can reduce damages, although the cost and effectiveness also vary. Finally, for fuel reduction activities on federal lands, delays and modifications — related to endangered species concerns and public involvement in decision-making — can affect the cost of fuel treatments.

Wildland Ecosystems and Wildfire

Ecosystem fire regimes can be classified in several ways; one common approach is to distinguish among surface fire ecosystems, stand-replacement fire ecosystems, and mixed fire ecosystems.[29] Damages to lands and resources depend on the nature of those ecosystems.

Surface-Fire Ecosystems

Surface-fire ecosystems are ecosystems where fires burned relatively frequently (typically 5- to 35-year intervals), with the fires consuming leaves or needles, grasses, twigs and small branches, and sometimes small trees, but generally leaving moderate and large trees unharmed by the fire. The classic surface-fire ecosystem is the western Ponderosa pine, where seedlings occasionally survive the surface fire to become the scattered, stately pines in fields of grass or low brush. The other archetypical surface fire ecosystem is the southern yellow pines — shortleaf, slash, loblolly, and especially longleaf pine. Surface-fire ecosystems account for about 34% of all U.S. wildlands.[30]

Over the past century, surface-fire ecosystems in the West have been affected by grazing, logging, and fire protection. Heavy grazing reduced grass cover, which commonly carried the surface fires. Logging in many areas emphasized the large pines, often leaving the true firs and Douglas-fir (which are more susceptible to drought, insect damage, and crown fires) to replace the pines, at least in the northern Rockies and Pacific Northwest. Fire protection has similarly led to more firs and Douglas-firs, and small Ponderosa pines, than would typically have survived. With fire return intervals of 5-35 years (i.e., fires typically burning once in that period), many surface-fire ecosystems have missed two or more burning cycles. Thus, many forests now have an unnaturally large accumulation of small burnable materials and of trees susceptible to crown fires.

Many are concerned that the unnatural fuel accumulations and the *fuel ladders* (continuous fuels from the ground to the tree crowns) from many small and medium- sized pines, firs, and Douglas-firs are causing crown fires in ecosystems where such fires were rare. This could cause significant ecological damage to plants and animals ill-adapted to crown fires. It is unclear whether a new surface-fire ecosystem will develop in the wake of an intense crown fire.

Research on fuel reduction treatments (discussed below) have documented the effectiveness of such treatments on Ponderosa pine (a surface-fire ecosystem),[31] and activities that reduce fuel accumulations have been shown to reduce wildfire severity in surface-fire ecosystems.[32] Presumably, less severe wildfires cause less damage to timber, to watersheds, and to wildlife and wildlife habitats.

Stand-Replacement-Fire/Crown-Fire Ecosystems

Stand-replacementfire ecosystems are ecosystems where crown fires are normal, natural, periodic events to which the ecosystem has adapted. The interval for the stand-replacement fires varies widely — from a few years (prairie grasses) to more than 1,000 years (coastal Douglas-fir) — depending on the ecosystem. Some ecosystems require periodic crown fires to regenerate the ecosystem. For example, lodgepole pine in much of the West and jack pine in the Lake States have *serotinous* cones, which only open and release their seeds after exposure to temperatures exceeding 250° Fahrenheit. Similarly, chaparral in southern California and the desert Southwest, most perennial grasses, and aspen everywhere regenerate from rootstocks; burning the surface vegetation allows new plants to sprout from the underground stems, rhizomes, and root crowns. Stand-replacement-fire ecosystems account for about 42% of all U.S. wildlands.[33]

It seems unlikely that stand-replacement-fire ecosystems could suffer significant ecological damage from severe wildfires. In contrast to surface-fire ecosystems, where crown fires could alter the ecosystem, in stand-replacement-fire ecosystems, the *exclusion* of crown fires (if it were possible) would likely alter the ecosystems. This ecological change is implied by evidence from grass ecosystems (prairies and meadows), where fire suppression is feasible and which are being encroached upon by trees that would normally have been eliminated by the frequent fires.

Activities that reduce fuel levels in stand-replacement-fire ecosystems have no documented effect on wildfire severity. Anecdotal reports have asserted that crown fires were halted (became surface fires) when they arrived at treated areas, but research has not documented where and when such occurrences have happened. To date, no research has shown that fuel treatments consistently reduce the extent or severity of wildfires in stand-replacement-fire ecosystems. The ineffectiveness of fuel reduction was particularly noted for southern California chaparral: "large fires were not dependent on old age classes of fuels, and it is thus unlikely that age class manipulation of fuels can prevent large fires."[34]

Mixed-Fire-Intensity Ecosystems

Many wildlands have ecosystems that burn in crown fires of relatively limited scale, substantially mixed with surface fires. These ecosystems are called mixed-fire-intensity ecosystems. A classic example is whitebark pine, a species generally limited to high elevation sites, near timberline. Whitebark pine is a slow-growing species that invades harsh sites and moderates the micro-climatic conditions to allow true firs and spruces to germinate and grow. The sporadic mixed-intensity fires kill most of the competing trees and some of the whitebark pines, but some pines survive. Also, burned sites are preferred "cache" sites for Clark's nutcrackers, which is the primary means of whitebark pine tree regeneration.[35] Other species that are commonly surface-fire or stand-replacementfire species, such as Ponderosa pine and lodgepole pine, can be mixed-fire-intensity types under certain conditions, typically near the transition to another area with a different dominant tree species. Ponderosa pine, for example, may be a mixed-fireintensity type on relatively moist sites, especially where it mixes naturally with Douglas-fir, such as on north-facing slopes in the northern Rockies. Lodgepole pine may be a mixed-fire-intensity type on relatively dry sites, where the trees naturally grow farther apart, such as on the eastern slopes of the Sierra Nevada Mountains.

Less is known about wildfire in mixed-fire-intensity ecosystems, even though they occupy about 24% of U.S. wildlands.[36] It is unclear whether fuel loads have accumulated to unnatural levels, whether crown fires could cause significant ecological damage, or whether fuel reduction activities would alter wildfire extent or severity in these ecosystems.

Wildfire Effects

The effects of wildfires on natural resources are difficult to assess and are commonly overstated for two reasons. First, burned areas look bad — blackened trees and ground cover — even following surface fires. However, many plants recover from being burned. Conifers generally survive even with as much as 60% of their crowns scorched.[37] Other plants, especially grasses, aspen, and some brush species, resprout vigorously after being burned. Furthermore, animals (regardless of their size and mobility) are only rarely killed by wildfire.[38]

The other reason that wildfire effects are commonly overstated is that reported burned area includes all the acres within the fire perimeter. However, even severe crown fires are patchy, leaving some areas lightly burned or unburned. For example, in the Yellowstone fires that were on the nightly news for weeks in the summer of 1988, 30% of the reported burned area was actually unburned and another 15-20% had only surface fire.[39] In the 2002 Hayman Fire, the worst wildfire in Colorado history, 35% of the area had a high severity burn and 16% had a moderate severity burn; 34% had a low severity burn and 15% was unburned.[40] Thus, severely burned acreage is substantially less than the burned area that is reported.

Severe wildfires can cause long-lasting resource damages. Crown fires kill many plants within the burned area, increasing the potential for erosion until the vegetation recovers. Some observers have reported "soil glassification," where the silica in the soils has been melted and fused, forming an impermeable layer in the soil, although research has yet to document the extent, frequency, and duration of the condition and the soils and conditions under which it occurs. Landslides can also occur in areas with unstable soils where the vegetation has burned, such as in coastal southern California. Timber can also be damaged, although burned trees can often be salvaged for lumber and other wood products. However, harvesting and processing costs are typically higher in burned areas, and many object to post-fire salvage harvesting because of its possible additional impacts on soils and other resource values. Wildfires, especially crown fires, can also have significant local economic effects — directly on tourism, and indirectly through effects on timber supply, water quality, and aesthetics. On the other hand, federal wildfire suppression efforts include substantial expenditures, many of which are made locally, and firefighting jobs are considered financially desirable in many areas.[41]

Protecting Wildlands and Resources

The federal government is generally responsible for protecting federal lands and their natural resources from wildfire.[42] Wildfire protection of other wildlands and natural resources — state, local government, and private lands — is the responsibility of the states, although the individual landowners are responsible for excessive fuel accumulations and other hazardous conditions on their own lands. As noted above, the FS has a technical and financial assistance program for state fire agencies.

The principal goal for land and resource protection is to reduce the damages caused by wildfires. This can best be achieved by reducing burnable biomass (live and dead) to reduce wildfire intensity and duration, and especially by eliminating the *fuel ladders* (relatively continuous biomass from the surface to tree crowns) that facilitate wildfire transition from a surface fire to a crown fire.[43] Fuel treatments can also reduce the *crown bulk density* (the biomass, especially fine fuels, in the tree crowns), making it more difficult for a crown fire to sustain itself, thus making a wildfire more controllable.[44] Reducing burnable biomass, however, does not eliminate wildfires, because fuel reduction does not directly alter the dryness of the biomass or the probability of an ignition.

The two principal mechanisms for reducing fuels are prescribed burning and mechanical treatments, although the two tools can also be combined. Each tool has benefits, costs, and risks or limitations to its use.

Prescribed Burning

Prescribed burning is intentionally setting fires in specified areas when fuel and weather conditions are within prescribed limits (e.g., fuel moisture content, relative humidity, wind speed). Some observers include, in their definition of prescribed burning, naturally occurring fires that are allowed to burn because they are within acceptable areas and conditions, as identified in fire management plans. The agencies term such fires *wildland fire use*, and do not identify them as prescribed fires, but do include the acres burned in wildland-fire-use fires as acres treated for fuel reduction.

Prescribed burning is used for reducing biomass fuels because it is the only means available for eliminating *fine fuels* (grasses, needles, leaves, forbs, and twigs and shrubs less than a quarter-inch in diameter [pencil-sized]). Burning converts the vegetation to smoke (carbon dioxide, water vapor, fine particulates, and other pollutants) and ashes (mineralized forms of the organic matter, readily available for absorption by new plant growth). Reducing fine fuels is critical in wildfire protection and control, because fine fuels are necessary to carry wildfires; without fine fuels, wildfires cannot spread.

Prescribed burning has various limitations, as well. Smoke can be a problem, contributing to human health problems, especially in areas where inversions are common or with relatively stagnant airsheds. Also, prescribed burning is risky. It is not *controlled* burning; there is no such thing as controlled burning, because there is no switch to turn the fire off. Prescribed fire is also an indiscriminate tool for reducing tree density, crown density, and fuel ladders, burning what is available, depending on a host of site-specific and micro-climatic conditions.

Finally, prescribed burning is expensive. Actually starting the prescribed fire is cheap — matches don't cost a lot. However, minimizing the risk to surrounding areas (especially private lands and housing developments) requires planning and preparation as well as having sufficient trained personnel and supervisors to react if (when) unexpected fire behavior occurs or weather conditions change. A prescribed fire that becomes a wildfire, such as the Cerro Grande Fire in Los Alamos, NM (that burned 237 houses in town) raises questions about the practice and about the fire managers who use it. Thus, fire managers tend to err on the side of excessive personnel (and cost) for a prescribed fire, rather than risk it becoming a costly, damaging wildfire with far higher costs.

Mechanical Treatment

Mechanical fuel treatment includes a wide array of activities designed to reduce biomass on a site. Foresters have a variety of terms for the various activities, including:[45]

- pruning — removing lower tree branches, which eliminates fuel ladders and can reduce crown density;
- release — removing several to many trees from a young stand (saplings or smaller) to concentrate wood growth on desirable trees, which reduces crown density;
- thinning — removing a portion of the standing trees; the portion can vary widely from very light (relatively few trees) to very heavy (more than half the trees in the stand). Thinning can be commercial (if the trees are large enough for products) or precommercial. It can be used to eliminate fuel ladders and reduce crown density, depending on the approach and portion of trees removed. Thinning approaches include:
  - low thinning, or thinning from below, to remove the smallest and poorest specimens, which eliminates fuel ladders and can reduce crown density;
  - crown thinning, or thinning from above, to open the canopy to stimulate growth on the remaining trees, which substantially reduces crown density;
  - selection thinning, to remove the least desirable trees for the future stand, which reduces crown density and can eliminate fuel ladders; and
  - mechanical thinning, to provide appropriate spacing for the remaining trees, which reduces crown density and can eliminate fuel ladders;
- salvage harvesting — removing a portion to all of the standing trees, many of which have been killed or are in imminent danger; includes presalvage harvesting (removing highly vulnerable trees before they are killed) and sanitation harvesting (removing trees to control the spread of insects or diseases). It reduces (or eliminates) crown bulk density, and might reduce fuel ladders.

Treatment Choices

Mechanical fuel treatment clearly involves choices — about the amount of biomass to be removed, and about the nature of the biomass to be removed (small and weak trees, lower limbs, vulnerable trees or species, etc.). The choice can also be over the method used for the treatment: a commercial sale, if the treatment yields commercially usable wood; a stewardship contract, if commercially usable wood can be exchanged for other activities; a service contract, for specified actions; an end-results contract, to specify what is left after treatment; or even treatment by agency personnel. All of these choices affect the public acceptance of the proposed treatment.

Benefits and Limitations

The primary benefit of mechanical fuel treatment is the high degree of control over the results. One report stated:[46]

> Mechanical thinning has the ability to more precisely create targeted stand structure than does prescribed fire ... Used alone, mechanical thinning, especially emphasizing the smaller trees and shrubs, can be effective in reducing the vertical fuel continuity that fosters initiation of crown fires. In addition, thinning of small material and pruning branches are more precise methods than prescribed fire for targeting ladder fuels and specific fuel components ...

The authors also observed some of the limitations of mechanical fuel treatment:[47]

> However, by itself mechanical thinning does little to beneficially affect surface fuels with the exception of possibly compacting, crushing, or masticating it during the thinning process. Depending on how it is accomplished, mechanical thinning may add to surface fuels (and increase surface fire intensity) unless the fine fuels that result from the thinning are removed from the stand or otherwise treated....
>
> Thinning and prescribed fires can modify understory microclimate that was previously buffered by overstory vegetation ... Thinned stands (open tree canopies) allow solar radiation to penetrate to the forest floor, which then increases surface temperatures, decreases fire fuel moisture, and decreases relative humidity compared to unthinned stands — conditions that can increase surface fire intensity ... An increase in surface fire intensity may increase the likelihood that overstory tree crowns ignite ...

Other sources have similarly reported the limitations of thinning:[48]

> Depending on the forest type and its structure, thinning has both positive and negative impacts on crown fire potential. Crown bulk density, surface fuel , and crown base height [fuel ladders] are primary stand characteristics that determine crown fire potential. Thinning from below, free thinning, and reserve tree shelterwoods have the greatest opportunity for reducing the risk of crown fire behavior. Selection thinning and crown thinning that maintain multiple crown layers ... will not reduce the risk of crown fires except in the driest ponderosa pine ... forests. Moreover, unless the surface fuels created by using these treatments are themselves treated, intense surface wildfire may result, likely negating positive effects of reducing crown fire potential. No single thinning approach can be applied to reduce the risk of wildfires in the multiple forest types of the West.

Thus, thinning and pruning have the potential to reduce the risk of crown fire, but may *increase* wildfire risk until the *slash* (non-commercial biomass) degrades (rots or burns, typically in a few years to decades, depending on the ecosystem), or is removed. In addition, thinning is an expensive proposition, with treatment costs ranging "from $35 to over $1000 per acre depending on the type of operation, terrain, and number of trees to be treated."[49]

Commercial operations — commercial thinning, stewardship contracting, and salvage logging — have been suggested as a means to moderate the high cost of mechanical fuel treatment. However, commercial timber sales on federal lands commonly cost more to prepare and administer than they return to the Treasury.[50] The results of commercial operations for fuel reduction are also questionable:[51]

> The proposal that commercial logging can reduce the incidence of canopy fires was untested in the scientific literature. Commercial logging focuses on large diameter trees and does not address crown base height — the branches, seedlings and saplings which contribute so significantly to the "ladder effect" in wildfire behavior. (Emphasis in original)

Others have also noted the likely net cost of thinning to reduce the risk of crown fires :52

> Although large trees can be removed for valuable products, the market value for the smaller logs may be less than the harvest and hauling charges, resulting in a net cost for thinning operations. However, the failure to remove these small logs results in the retention of ladder fuels that support crown fires with destructive impacts to the forest landscape. A cost/benefit analysis broadened to include market and nonmarket considerations indicates that the negative impacts of crown fires are underestimated and that the benefits of government investments in fuel reductions are substantial.

Combined Operations

The ability to control the resulting stand structure with mechanical treatments and the ability to remove fine fuels with prescribed burning make combining the two treatments seem a logical choice. However, empirical evidence to document the effectiveness of such combined operations is limited:[53]

> A more limited number of studies addressed the effectiveness of a combination of thinning and burning in moderating wildfire behavior. The impacts varied, depending on the treatment of the thinning slash prior to burning.... (Emphasis in original)

In addition, the cost of combined operations is substantially greater than the cost of either alone.

Area Needing Treatment

The areas that might benefit from prescribed burning and/or mechanical treatment is not entirely clear. Table 1, below, shows the acreage of national forest land, Department of the Interior land, and all other land by historical fire regime (comparable to the ecosystem types described above); and condition class — low risk (Class 1), moderate risk (Class 2), and high risk (Class 3) of losing key ecosystem components in a wildfire.

Based on the discussion (above) of the effectiveness of various treatments, it seems reasonable to conclude that treating lands in class 3 (high risk), low severity (surface fire) regime could reduce the likelihood of crown fires in these ecosystems where such fires are unnatural (or at least very rare). **Table 1** shows this to include 28.8 million acres of national forest land, 6.5 million acres of Interior land, and 42.2 million acres of other federal, state, and private land.

The cost to treat these lands varies widely. One study, cited above, reported mechanical treatment costs of $35 to $1,000 per acre, depending on terrain, type of operation, and number of trees to be cut.[54] Others have similarly reported highly variable costs for commercial mechanical treatment above and below the "base case" cost of $150 per acre, depending on tree size, stand density, terrain, and whether the treatment was conducted in the wildland-urban interface.[55] The same source reported similar variability in costs for prescribed burning, above and below the "base case" cost of $ 105 per acre. Federal appropriations for fuel treatment averaged about $170 per acre for FY200 1 — FY2006 — $165 per acre for the Forest Service and $174 per acre for the BLM.[56] The General Accounting Office (GAO, now the Government Accountability Office) used a Forest Service estimate of $300 per acre in its 1999 estimate of needed funding for fuel treatment, because of the higher cost per acre to treat additional western lands.[57] At $300 per acre, Forest Service costs to treat the Class 3 — surface fire regime lands would be $8.6 billion, and Department of the Interior costs would be $1.9 billion. Other surface (low severity) fire regime lands might also warrant treatment, although the lower risk of ecological damage suggests a lower priority for treatment.

Table 1. Lands At Risk of Ecological Damage from Wildfire, by Landowner Group and Historical Fire Regime in millions of acres

Landowner/ Historical Fire Regime	Total	Class 1 (low risk)	Class 2 (mod. risk)	Class 3 (high risk)
USDA Forest Service				
Low severity (surface fire)	83.67	19.87	34.96	28.83
Mixed severity	53.93	16.05	26.71	11.17
Stand replacement	58.93	29.03	18.77	11.13
Forest Service, total	196.52	64.95	80.45	51.12
Dept. of the Interior				
Low severity (surface fire)	49.00	18.70	23.83	6.46
Mixed severity	97.80	62.05	25.82	9.92
Stand replacement	80.93	47.67	26.17	7.09
Interior Dept., total	227.72	128.42	75.83	23.47
Other federal, state, and private lands				
Low severity (surface fire)	296.02	136.46	117.37	42.20
Mixed severity	142.18	49.55	59.72	32.92
Stand replacement	386.81	217.46	137.28	32.07
Other lands, total	825.01	404.60	313.24	107.18

Source: Kirsten M. Schmidt, James P. Menakis, Colin C. Hardy, Wendel J. Hann, and David L. Bunnell, *Development of Coarse-Scale Spatial Data for Wildland Fire and Fuel Management*, Gen. Tech. Rept. RMRS-87 (Ft. Collins, CO: USDA Forest Service, Rocky Mountain Research Station, Apr. 2002), pp. 13-15.

It is unclear whether any lands other than the surface fire regime lands warrant fuel treatment. The existing research evidence on fuel treatment for stand replacement fire regimes raises questions about the effectiveness of both mechanical treatment and prescribed fire for reducing the likelihood of damages from a crown fire. One might even question whether ecological damage can be ascribed to a crown fire in a stand-replacement fire ecosystem, since these ecosystems have evolved adaptations to reestablish themselves following crown fires. Evidence is also lacking on the effectiveness of mechanical treatments and prescribed burning on mixed- intensity fire ecosystems. Thus, it is not certain whether fuel treatment on these mixed-intensity fire regime lands and stand-replacement fire regime lands would provide any significant wildfire protection.

Delays and Changes in Federal Decision-Making

Some advocates of fuel treatment are concerned that the delays and changes to the implementation of fuel treatments might lead to catastrophic crown fires that could have been prevented by more expeditious fuel treatment. Concerns are generally linked to consultations under the Endangered Species Act (ESA, P.L. 93- 205; 16 U.S.C. §§ 1531-1544), and to public involvement under the National Environmental Policy Act of 1969 (NEPA; P.L. 91-190, 42 U.S.C. §§ 4321-4347) and the Forest Service Appeals Reform Act (ARA; §322 of P.L. 102-381 [the FY1993 Interior appropriations act], 16 U.S.C. § 1612 note).[58]

Involving the public and consulting over possible impacts on endangered or threatened species take time, and concerns and objections can delay, modify, or even prevent some proposed actions. However, others caution that expedited review or limits on ESA consultation and on public oversight of proposed fuel treatments may allow treatments to include commercial timber harvests or other actions that provide little wildfire protection and exacerbate fuel accumulations in the short run, while causing other environmental damages.

This raises the question of the effect of delays on wildfire threats. Clearly, structures in the wildland-urban interface are threatened by wildfire, but as shown above, fuel treatment provides little, if any, fire protection for structures, and thus delaying fuel treatments has little consequence for structure protection. Resources in surface-fire ecosystems with unnatural fuel accumulations are at risk from severe wildfires. The odds of having treated the "right" acres to prevent a crown fire with significant resource damages are, however, quite low. Over the past four years, during which more area burned annually than in any other year since 1960, wildfires have burned an average of 9.0 million acres annually. Total wildlands in the United States are 1.47 *billion* acres — 653.3 million acres of federal land,[59] and 812.1 million acres of private forest and rangeland.[60] Thus, the likelihood of any particular acre burning in any given year, on average, is less than 0.66% (i.e., burning once every 150 years). Obviously, the risk for certain areas in particular years can be much higher — 5.4% of Idaho's wildlands burned in 2007, for example — but this is offset by much lower risks for those areas in other years and for other areas in the same year — 0.2% of Idaho's wildlands burned in 2002, while 0.04% of Colorado wildlands burned in 2007, in contrast to 1.8% in 2002, when the Hayman Fire burned.[61] Wildfire risk is probably somewhat higher in western states than the national average, because the ecosystems in the Lake States, Mid-Atlantic region, and New England experience less fire; however, even if the risk were 50% greater than the national average (which seems unlikely because the larger area in the West already contributes to a higher national average), the risk would still be less than 1% per year.

In addition to the low probability of a particular acre burning is the modest likelihood of an area being treated. The Forest Service and BLM have treated 2.7 million acres of their lands annually over the past five years.[62] This is less than 8% of their Class 3 — surface fire ecosystem lands, and less than 3% of Class 3 plus Class 2 surface fire ecosystem lands. If the same acreage of treatments are spread more broadly — to Class 1 — surface fire ecosystem lands or to lands in other fire regimes — the probability of treating a particular acre to prevent a crown fire diminishes further.

Nonetheless, lengthy delays can exacerbate the risks. Annual probabilities of a wildfire burning an area and of an area being treated are both cumulative. Over a 10-year period, the likelihood of an area burning is more than 6%, while the likelihood of a moderate- or high-risk surface-fire ecosystem being treated rises to 15% (if half of all treatments are concentrated on these lands). Thus, relatively brief delays may have relatively little impact of the likelihood of an area being burned in an unnatural crown fire, but longer delays (a decade or more) could have a significant impact.

ESA Consultations.[63]

The ESA established a process for federal agencies to consult with the Fish and Wildlife Service (FWS), or with the National Marine Fisheries Service (NMFS) for some species, on any actions that *might* jeopardize a listed endangered or threatened species or adversely modify its critical habitat. This is not a problem for fire-fighting, as immediate, informal consultations can occur during an emergency, with formal consultation to follow after the emergency has passed. However, some fuel treatments might jeopardize a species or adversely modify its habitat, which would require ESA consultation. Consultation means the FWS (or NMFS) would review the proposed action and, if jeopardy or adverse habitat impacts are likely, propose a "reasonable and prudent alternative" to achieve the same purpose without jeopardy or adverse habitat modification. The vast majority of agency activities have a finding of no jeopardy, and most with jeopardy have a reasonable and prudent alternative; actions with jeopardy and no alternative findings are exceedingly rare.

Fuel treatments that reduce the likelihood of crown fires in ecosystems where such fires were historically rare are generally unlikely to jeopardize or adversely modify the critical habitat of endangered species. Many species in North America are adapted to survive and even thrive with natural wildfires. One study reported that more than 90% of rare, threatened, and endangered plants in the 48 coterminous states either benefit from fire or are found in fire-adapted ecosystems.[64] Also, as noted above, animal mortality in wildfires is rare. Thus, treatments that only restore forests to conditions that allow an historically natural ecological role for wildfire are more likely to benefit endangered and threatened species than to harm them.

Nonetheless, ESA consultations take time, and can delay fuel treatments. This is more likely to be the case when restoration treatments (e.g., prescribed burning or thinning from below) are combined with other activities (e.g., commercial timber harvesting), such as in a stewardship contract. Thus, the method used to undertake the treatment, as well as the nature of the treatment itself, determines the length of delays and possible project modifications from ESA consultations.

NEPA Environmental Analysis and Public Involvement.[65]

NEPA requires federal agencies to review the environmental effects of "major Federal actions significantly affecting the quality of the human environment." Agencies must consider every significant aspect of the environmental impacts of a proposed action before making an irreversible commitment of resources to the project. NEPA also requires that agencies inform the public that they have considered those impacts in their decision-making process. In his Executive Order on NEPA implementation, President Richard Nixon directed the agencies to go beyond just informing the public, to actively involve the public early in the decision-making process.[66] Fuel reduction treatments to protect resources from wildfires are generally considered to be major federal actions subject to NEPA.

Environmental Analysis

The action agency must analyze the possible environmental consequences of its actions. The first step is to determine if the action will have significant environmental impacts. There are three possible outcomes. If significant impacts are likely, then the agency prepares an environmental impact statement (EIS). If the impacts are normally insignificant — individually and cumulatively — the activity can be *categorically excluded* from further NEPA environmental analysis and public involvement. (See below.) If the significance of the impacts is uncertain, the agency prepares an environmental assessment (EA) to determine the significance of the impacts. The EA leads either to a finding of no significant impact (FONSI) or to an EIS.

Advocates of expedited fuel treatment are concerned about the time needed to prepare an EIS or even an EA. Information collection and analysis may take from several days to a few months, depending on the magnitude and complexity of the proposed action. An EIS involves additional steps to assess the likely and the possible environmental impacts and to inform and involve the public. These steps include *scoping* (public discussions about the nature, location, and possible consequences of the proposal); a draft EIS, examining a range of alternatives and generally identifying a preferred alternative; public comments on the draft and the preferred alternative; and then a final EIS and Record of Decision (ROD).[67] Only after completing this process — which can take a year or more for large, complex projects — can the agency undertake the action. Thus, proponents of expeditious fuel reduction projects often advocate various approaches to accelerate the process, discussed below.

Categorical Exclusions (CEs)

As noted above, certain projects can be categorically excluded from the requirement to prepare an EA or an EIS. Such a CE action is defined as:[68]

> ... a category of actions which do not individually or cumulatively have a significant effect on the human environment ... and for which, therefore, neither an environmental assessment nor an environmental impact statement is required.... Any procedures under this section shall provide for extraordinary circumstances in which a normally excluded action may have a significant environmental effect.

CEs are typically used for relatively minor, routine actions that the agency does frequently and has found to have at most insignificant environmental impacts. For projects approved under CEs, the Forest Service is not required to provide notice and opportunity for public comment as otherwise required for agency activities under the ARA. (See below.)

In certain situations — such as controversial issues (e.g., wetlands and roadless areas) or specifically protected resources (e.g., endangered species and archaeological sites) — known as *extraordinary circumstances*, CEs cannot be used. In 2002, the Forest Service modified its application of extraordinary circumstances, allowing the responsible official to determine whether the extraordinary circumstances warranted an EA or an EIS rather than automatically precluding use of a CE in the presence of extraordinary circumstances.[69]

The Forest Service has identified numerous categories of actions for which a CE may be used; two relate directly to wildfire protection (for details on Forest Service CEs and extraordinary circumstances, see Appendix A):[70]

6. *Timber stand and/or wildlife habitat improvement activities* ..., [including] thinning or brush control to improve growth or reduce fire hazard ..., prescribed burning to control understory hardwoods in stands of southern pine, and prescribed burning to reduce natural fuel build-up ....

10. *Hazardous fuel reduction activities using prescribed fire, not to exceed 4,500 acres, and mechanical methods for crushing, piling, thinning, pruning, cutting, chipping, mulching, and mowing, not to exceed 1,000 acres* ... limited to ... the wildland-urban interface; or Condition Classes 2 or 3 [moderate or high risk of ecological damage] in Fire Regimes I, II, or III [surface fire, stand- replacement fire with a return interval of 35 years or less, and mixed-intensity fire] ... (Emphasis in original) Forest Service use of the latter CE was halted after a court found it was arbitrary and capricious.[71] Other CEs have also been challenged, raising questions about the availability of CEs for fuel reduction projects.[72]

Forest Service Appeals Reform Act

In addition to public involvement under NEPA, the Forest Service must also inform the public of its decisions and provide an opportunity for the public to request an administrative review of its decisions under the Forest Service Decisionmaking and Appeals Reform Act (ARA).[73] Subsections (a) and (b) require the Forest Service to provide notice and an opportunity for public comment on proposed actions; this is the only provision requiring notice and comment on Forest Service proposals other than under NEPA. Subsections (c) and (d) specify an administrative appeals process — review by a higher-ranking official — for those who had commented on the proposal and object to the decision.

GAO was asked to examine administrative appeals of fuel reduction projects.[74] For FY2001 and FY2002, prior to promulgation of the hazardous fuel reduction CE, 59% of fuel reduction projects used CEs and could not be appealed. Of those that could be appealed, 58% were appealed (i.e., 24% of all fuel reduction projects during that period). Of those, 73% were implemented without change, 8% were modified, and 19% (less than 5% of all projects) were withdrawn or reversed. Furthermore, 79% of the appeals were resolved within the prescribed 90 days. These data are supported by a study of all Forest Service administrative appeals.[75] This study found that 8% of appeals were granted (i.e., decision reversed) and that 9% of appealed decisions were withdrawn.

A different study examined factors that increased the likelihood of a fuel reduction project being appealed.[76] They reported that appeals were more likely for fuel reduction projects that (1) affected more area; (2) included more activities for the site; (3) included commercial timber harvest; (4) included as a purpose reducing fuels generated by the project; and (5) had at least one threatened or endangered mammal near the site. These factors are indirectly confirmed in the GAO study, since 92% of projects with EISs (larger projects with likely environmental impacts) were appealed, compared to 52% of projects with EAs (projects with uncertain environmental impacts).[77] Conversely, projects were significantly less likely to be appealed if the project was: (1) implemented by Forest Service personnel or a service contract; and (2) in the wildland-urban interface.

These data suggest that administrative appeals are less of a problem than the advocates of fuel treatment suggests. Only about a quarter of proposed projects are appealed, with less than 5% prevented from being implemented, and delays of less than 90 days for most projects. However, for prescribed burning, a 90-day delay can be significant, since the period within the prescribed conditions can be brief.

Expedited Procedures

Proponents of aggressive fuel treatment continue to be concerned about delays from the ESA, NEPA, and ARA review processes, and have pressed for various means for accelerating the reviews. Some procedures are currently feasible under existing regulations, others have been enacted by Congress in various contexts, and more have been proposed.

Expedited ESA Consultations

As noted above, during emergencies, the agencies can consult informally for rapid action, with formal consultations to follow when the situation has stabilized. This clearly applied during wildfire suppression activities, but fuel reduction treatments are not emergency actions that require an immediate response to prevent damages. As discussed above, lengthy (multi-year) delays in fuel reduction activities can increase the likelihood of resource damages from wildfires, but brief delays have minor impacts.

The agencies have developed an alternative approach to ESA consultations that is intended to accelerate the ESA review process: *counterpart* regulations.[78] These regulations allow the Forest Service, BLM, and others to assess whether the proposed fuel reduction action is likely to jeopardize a listed threatened or endangered species or to adversely modify critical habitat, rather than to consult with the Fish and Wildfire Service on the likelihood of jeopardy or adverse habitat modification. While some ESA counterpart regulations have been challenged successfully, the counterpart regulations related to wildfire management remain in place.[79]

Expedited NEPA Reviews (other than through CEs)

In addition to the option of CEs, the NEPA regulations of the Council on Environmental Quality (CEQ) allow for *alternative arrangements* in the event of an emergency.[80] These alternative arrangements do not waive NEPA requirements, but establish an alternative means of fulfilling those requirements for actions necessary to control the immediate impacts of an emergency, typically with conditions on short-term and long-term actions.[81] For example, in 1998, the Forest Service requested alternative arrangements for rapid restoration actions following a windstorm that damaged 103,000 acres of national forest land in Texas that contained critical habitat for the endangered red-cockaded woodpecker; CEQ concurred that the situation was an emergency and agreed to alternative arrangements that included subsequent preparation of an EA, limits on tree removal, long-term public involvement, emergency consultation under ESA, and more.

For fuel treatment, NEPA alternative arrangements will rarely provide a means of accelerated action. First, alternative arrangements are not used very often — 41 requests were made from 1978 through 2006.[82] Second, alternative arrangements are to be used for emergencies. Fuel conditions in a delineated area might occasionally be an emergency, such as in the wake of a ice storm or a tornado, but fuel levels generally do not constitute an emergency requiring immediate action.

Healthy Forests Restoration Act

The Healthy Forests Restoration Act of 2003 (HFRA; P.L. 108-148, 16 U.S.C. §§ 6501-659 1) expedited review processes in several ways. In Title I, it modifies the NEPA environmental analysis and public involvement processes for authorized Forest Service and

BLM fuel reduction projects (based on priorities, exclusions, and other standards in the act). The EA or EIS for each project may be limited to the proposed action, the no-action alternative, and possibly an additional alternative (in contrast to the range of alternatives normally required). The agencies "shall facilitate collaboration" with tribes and state and local governments and "participation" of interested persons; however, it does not explain the distinction between *collaboration* with certain interests and *participation* by other interests.

Title I includes two other changes to accelerate fuel reduction projects. First, for the Forest Service, it replaces ARA administrative appeals with a "predecisional administrative review process." This process is only available to persons who submitted "specific written comments that relate to the proposed action" during scoping or the public comment period on the draft NEPA document. The process is also limited to the period between completing the EA or EIS and issuing the Record of Decision, with no requirements for how long that period must be. Then, the act restricts judicial review, generally limiting plaintiffs to those who have exhausted administrative review processes and specifying the venue for review, while encouraging expeditious judicial review and requiring the courts to balance the short- and long-term effects of action and inaction in deciding on injunctions.

In Title IV, HFRA allows the use of CEs for "applied silvicultural assessments" — timber harvesting and other vegetative treatments "for information gathering and research purposes." Each treatment is limited to 1,000 acres, with exclusions for certain areas and limitations on the adjacency of treatments, and with public notice and comment and "peer reviewed by scientific experts selected by the Secretary [of Agriculture or of the Interior], which shall include non-Federal experts." Total acreage of all applied silvicultural assessments using this CE is limited to 250,000 acres.

Other Possibilities

Congress can create other means of accelerating the decision-making process for fuel reduction treatments. Congress has exempted certain federal activities (such as construction of the Trans-Alaska Pipeline to deliver oil from the North Slope) from NEPA compliance.[83] Congress has also directed in law that no EIS or EA be prepared in certain instances, through direct statutory language or by deeming that the authorized activities are not major federal actions that significantly affect the human environment. Congress has also pronounced certain analyses or substitute processes to be sufficient or adequate under NEPA.

Congress has also established alternative review processes. Typically this is in addition to NEPA public involvement, to accelerate the review by obtaining broader, organized review early in the decision-making process, vetting the decision before public review. Examples include resource advisory committees (RACs) under § 403 of the Federal Land Policy and Management Act of 1976 (FLPMA; P.L. 94-579, 43 U.S.C. § 1753) and under Title II of the Secure Rural Schools and Community Self- Determination Act of 2000 (P.L. 106-393; 16 U.S.C. § 500 note). Other advisory or collaborative groups have been established or acknowledged statutorily, commonly to provide supplemental public involvement.

Considerations in Expediting Decisions

Public acceptance of options to accelerate fuel treatments depends on a variety of factors. In general, earlier discourse among interests about the risks and needed treatments lead to greater comfort with the resulting decisions. One study found that survey respondents were willing to accept limitations on the rights to appeal and litigate agency decisions, but wanted to be more informed and involved in those decisions.[84] Greater specificity in approved treatments also is likely to result in greater acceptance. For example, a treatment prescription that specifies "thinning from below to approximately 20-foot spacing of remaining trees and emphasizing retention of Ponderosa pine" is likely to be more acceptable than "mechanical treatment to reduce stand density." Finally, authors have identified the need for *collective action* to minimize conflict over decisions, and three broad social factors to achieve collective action: developing collaborative capacity, framing problems in mutually-understood terms, and creating mutual trust among groups.[85] These are factors that take time, and cannot be legislated directly, although Congress can foster (or negate) their development by the ways in which it authorizes agency action to promote wildfire protection.

CONCLUSIONS

As more acres and more homes have burned in the past few years, and more people at risk from wildfires, Congress has faced increasing pressures to protect structures and resources. Congress decides what programs to authorize and fund, and many options exist.

To protect homes, Congress could create new programs and expand existing ones for installing non-flammable roofing, removing burnable materials adjacent to structures, and creating a defensible space of at least 30 feet around the building. Programs could inform homeowners, or assist or require landowner action; the programs could be federal or implemented through state or local governments.

Protecting resources poses different challenges for Congress, because ecological damages vary widely, depending on the ecosystem and on site-specific conditions. Fuel reduction can probably moderate crown fire damages in surface-fire ecosystems, and possibly in mixed-intensity-fire ecosystems. Existing programs for federal lands authorize prescribed burning (intentional fires under prescribed conditions) and mechanical treatments (cutting and removing some trees), the principal means of reducing fuel levels. However, prescribed fires are risky and mechanical treatments can cause other ecological damages, and both are expensive. Proponents of more fuel treatment advocate accelerated processes for environmental analysis and public review to reduce costs and expedite action. Others caution that inadequate analysis and review can allow projects with unintended damages and few fire protection benefits. Congress can alter the existing environmental and public review processes, recognizing the trade-offs between expeditious action and insufficient review. However, the fact is that crown fires occur; they cannot be halted and the damages they cause cannot be totally prevented.

APPENDIX A: EXCERPTS FROM FOREST SERVICE HANDBOOK ON NEPA CATEGORICAL EXCLUSIONS RELATED TO STRUCTURAL OR RESOURCE PROTECTION FROM WILDFIRES

The following material are excerpts from the Forest Service handbook on NEPA categorical exclusions — *FSH 1909.15 - Environmental Policy and Procedures Handbook. Chapter 30 - Categorical Exclusion from Documentation*, Amendment No. 1909.15-2007-1 (Feb. 15, 2007). Emphases (underscoring and boldface font) are in the original.

30.31. Policy

2. Resource conditions that should be considered in determining whether extraordinary circumstances related to the proposed action warrant further analysis and documentation in an EA or EIS are:
   a. Federally listed threatened or endangered species or designated critical habitat, species proposed for Federal listing or proposed critical habitat, or Forest Service sensitive species.
   b. Flood plains, wetlands, or municipal watersheds.
   c. Congressionally designated areas, such as wilderness, wilderness study areas, or national recreation areas.
   d. Inventoried roadless areas.
   e. Research natural areas.
   f. American Indians or Alaska Native religious or cultural sites.
   g. Archaeological sites, or historic properties or areas.

31.2 - Categories of Actions for Which a Project or Case File and Decision Memo Are Required

6. Timber stand and/or wildlife habitat improvement activities which do not include the use of herbicides or do not require more than one mile of low standard road construction .... Examples include but are not limited to:
   b. Thinning or brush control to improve growth or to reduce fire hazard including the opening of an existing road to a dense timber stand.
   c. Prescribed burning to control understory hardwoods in stands of southern pine.
   d. Prescribed burning to reduce natural fuel build-up and improve plant vigor.

10. Hazardous fuels reduction activities using prescribed fire, not to exceed 4,500 acres, and mechanical methods for crushing, piling, thinning, pruning, cutting, chipping, mulching, and mowing, not to exceed 1,000 acres. Such activities:
   a. Shall be limited to areas:
      (1) In the wildland-urban interface; or
      (2) Condition Classes 2 or 3 [moderate or high risk of ecological damage] in Fire Regimes I, II, or III [surface fire, stand-replacement fire at 35 years or less, and mixed-intensity fire], outside the wildland-urban interface;

b. Shall be identified through a collaborative framework as described in "A Collaborative Approach for Reducing Wildland Fire Risks to Communities and Environment 10-Year Comprehensive Strategy Implementation Plan";
c. Shall be conducted consistent with agency and Departmental procedures and applicable land and resource management plans;
d. Shall not be conducted in wilderness areas or impair the suitability of wilderness study areas for preservation as wilderness; and
e. Shall not include the use of herbicides or pesticides or the construction of new permanent roads or other new permanent infrastructure; and may include the sale of vegetative material if the primary purpose of the activity is hazardous fuel reduction.

...

12. *Harvest of live trees not to exceed 70 acres, requiring no more than 1/2 mile of temporary road construction.* Do not use this category for even-aged regeneration harvest or vegetation type conversion. The proposed action may include incidental removal of trees for landings, skid trails, and road clearing. Examples include but are not limited to:
    a. Removal of individual trees for sawlogs, specialty products, or fuelwood.
    b. Commercial thinning of overstocked stands to achieve the desired stocking level to increase health and vigor.
13. *Salvage of dead and/or dying trees not to exceed 250 acres, requiring no more than 1/2 mile of temporary road construction.* The proposed action may include incidental removal of live or dead trees for landings, skid trails, and road clearing. Examples include but are not limited to:
    a. Harvest of a portion of a stand damaged by a wind or ice event and construction of a short temporary road to access the damaged trees.
    b. Harvest of fire-damaged trees.
14. Commercial and non-commercial sanitation harvest of trees to control insects or disease not to exceed 250 acres, requiring no more than 1/2 mile of temporary road construction, including removal of infested/infected trees and adjacent live uninfested/uninfected trees as determined necessary to control the spread of insects or disease. The proposed action may include incidental removal of live or dead trees for landings, skid trails, and road clearing. Examples include but are not limited to:
    a. Felling and harvest of trees infested with southern pine beetles and immediately adjacent uninfested trees to control expanding spot infestations.
    b. Removal and/or destruction of infested trees affected by a new exotic insect or disease, such as emerald ash borer, Asian long horned beetle, and sudden oak death pathogen.

End Notes

[1] National Interagency Fire Center, "Fire Information — Wildland Fire Statistics," at [http://www.nifc.gov/fire_info/fires_stats.htm]. Fire season severity is commonly assessed by acres burned, but larger fires may not be "worse" if they burn less intensely, because their damages may be lower. However, fire intensity and damages are not measured for each wildfire, and thus cannot be used to gauge the severity of fire seasons. It is unclear whether acreage burned might be a reasonable approximation of severity.

[2] The past four fire seasons have surpassed the 2000 season, while the 1988 data have been revised downward.

[3] See Stephen F. Arno and Steven Allison-Bunnell, *Flames in Our Forest: Disaster or Renewal?* (Washington, DC: Island Press, 2002), pp. 45-46.

[4] For information on the interface, see CRS Report RS2 1880, *Wildfire Protection in the Wildland-Urban Interface*, by Ross W. Gorte.

[5] National Wildland/Urban Interface, Fire Protection Program, *Wildland/Urban Interface Fire Hazard Assessment Methodology*, p. 5, at [http://www.firewise.org/resources wham.pdf].

[6] Jack D. Cohen, "Preventing Disaster: Home Ignitability in the Wildland-Urban Interface," *Journal of Forestry* (March 2000): p. 17.

[7] *FireSmart: Protecting Your Community From Wildfire* (Edmonton, Alberta: Partners in Protection, May 1999), p. 2-5.

[8] *Wildland/Urban Interface Fire Hazard Assessment Methodology*, p. 7.

[9] *Is Your Home Protected From Wildfire Disaster? A Homeowner's Guide to Wildfire Retrofit* (Tampa, FL: Institute for Business and Home Safety, n.d.), p. 9, at [http://www. firewise.org/resources

[10] *Wildland/Urban Interface Fire Hazard Assessment Methodology*, p. 7.

[11] Jack Cohen, "The Cerro Grande Fire: Why Houses Burned," *Forest Trust Quarterly Report*, No. 23 (Dec. 2000): p. 7.

[12] *Wildland/Urban Interface Fire Hazard Assessment Methodology*, p. 7.

[13] *FireSmart*, p. 2-11, and Robert Bardon and Robin Carter, *Minimizing Wildfire Risk — A Forest Landowner's Guide*, 03/03 — 30M — DSB/SSS AG-616 (North Carolina Cooperative Extension Service, n.d.), p. 6.

[14] *Wildland/Urban Interface Fire Hazard Assessment Methodology*, p. 7.

[15] Jack D. Cohen, "Reducing the Wildland Fire Threat to Homes: Where and How Much?" *Proceedings of the Symposium on Fire Economics, Planning, and Policy: Bottom Lines*, Gen. Tech. Rept. PSW-GTR-173 (Berkeley, CA: USDA Forest Service, Pacific Southwest Research Station, Dec. 1999): pp. 189-195.

[16] Cohen, "Reducing the Wildland Fire Threat to Homes," p. 191.

[17] Cohen, "Reducing the Wildland Fire Threat to Homes," pp. 191-192.

[18] *FireSmart*, p. 2-16.

[19] Jack Cohen and Rick Stratton, "Home Destruction Within the Hayman Fire Perimeter," *Hayman Fire Case Study*, Gen. Tech. Rept. RMRS-GTR- 114 (Ft. Collins, CO: USDA Forest Service, Rocky Mountain Research Station, Sept. 2003): p. 264.

[20] Cohen and Stratton, "Home Destruction Within the Hayman Fire," p. 263.

[21] Peter Robichaud, Lee MacDonald, Jeff Freeouf, Dan Neary, Deborah Martin, and Louise Ashman, "Postfire Rehabilitation of the Hayman Fire," *Hayman Fire Case Study*, Gen. Tech. Rept. RMRS-GTR-1 14 (Ft. Collins, CO: USDA Forest Service, Rocky Mountain Research Station, Sept. 2003): p. 294.

[22] Model building codes for local governments exist. For example, in January 2006, the International Code Council (ICC) published its International Wildland-Urban Interface Code [http://www.iccsafe.org/news/about/#cd], with a separate chapter on ignition-resistant construction specifications. In June 2007, the National Fire Protection Association (NFPA) approved an updated standard, *Standard for Reducing Structure Ignition Hazards from Wildland Fire*, [http://www.nfpa.org/aboutthecodes/AboutTheCodes.asp?DocNum=1 144].

[23] See [http://www.bsc.ca.gov/default.htm].

[24] For a description of these programs, see CRS Report RL3 1065, *Forestry Assistance Programs*, by Ross W. Gorte.

[25] FIREWISE is part of the National Wildland/Urban Interface Fire Program, directed and sponsored by the Wildland/Urban Interface Working Team, which includes the USDA Forest Service, several DOI agencies (the Bureau of Land Management, Bureau of Indian Affairs, Fish and Wildlife Service, and National Park Service), and two agencies of the Department of Homeland Security (the Federal Emergency Management Agency and U.S. Fire Administration). It also includes the International Association of Fire Chiefs, National Association of State Fire Marshals, National Association of State Foresters, National Emergency Management Association, and National Fire Protection Association.

[26] See the program announced by the U.S. Fire Administration and the National Wildfire Coordinating Group, at [https://www.usfa.dhs.gov/media 1 2308.shtm].

[27] See the National Institute of Standards and Technology website, at [http://www2.bfrl. nist.gov/projects/2008ProgramContainer.asp?BFRLProgram=FWUI].

[28] For information on this program, see CRS Report RS22394, *National Flood Insurance Program: Treasury Borrowing in the Aftermath of Hurricane Katrina*, by Rawle O. King.

[29] Arno and Allison-Bunnell, *Flames in Our Forest*, pp. 44-45.

[30] Fire Modeling Institute, *Historical Fire Regimes by Current Condition Classes: Data Summary Tables* (Missoula, MT: USDA Forest Service, Rocky Mountain Research Station, Feb. 22, 2001), at [http://www.fs.fed.us/fire

[31] Russell T. Graham, Alan E. Harvey, Theresa B. Jain, and Jonalea R. Tonn, *The Effects of Thinning and Similar Stand Treatments on Fire Behavior in Western Forests*, Gen. Tech. Rept. PNW-GTR-463 (Portland, OR: USDA Forest Service, Pacific Northwest Research Station, 1999), 27 pp.

[32] Philip N. Omi and Erik J. Martinson, *Final Report: Effect of Fuels Treatment on Wildfire Severity*, submitted to the Joint Fire Science Program Governing Board (Ft. Collins, CO: Colorado State Univ., Western Forest Fire Research Center, Mar. 25, 2002).

[33] *Historical Fire Regimes by Current Condition Classes.*

[34] Jon E. Keeley, C.J. Fotheringham, and Marco Marais, "Reexamining Fire Suppression Impacts on Brushland Fire Regimes," *Science*, v. 284 (June 11, 1999): p. 1829.

[35] Diana F. Tomback, "Clark's Nutcracker; Agent of Regeneration," *Whitebark Pine Ecology and Restoration*, Diana F. Tomback, Stephen F. Arno, and Robert E. Keane, eds. (Washington, DC, Island Press, 2001), pp. 90-100.

[36] *Historical Fire Regimes by Current Condition Classes.*

[37] See Ross W. Gorte, *Fire Effects Appraisal: The Wisconsin DNR Example*, Ph.D. dissertation (East Lansing, MI: Michigan State Univ., June 1981).

[38] L. Jack Lyon, Mark H. Huff, Robert G. Hooper, Edmund S. Telfer, David Scott Schreiner, and Jane Kapler Smith, *Wildland Fire in Ecosystems: Effects of Fire on Fauna*, Gen. Tech. Rept. RMRS-GTR-42-vol. 1 (Ogden, UT: USDA Forest Service, Rocky Mountain Research Station, Jan. 2000).

[39] Lyon, et al., *Effects of Fire on Fauna*, p. 44.

[40] Robichaud et al., "Postfire Rehabilitation of the Hayman Fire," p. 294.

[41] Nelson, *A Burning Issue*, pp. 37-38.

[42] The federal government also protects some state and private lands where the landowner has a cooperative agreement with a federal agency, while some federal lands similarly are protected by state or private organizations under cooperative agreements.

[43] Russell T. Graham, Alan E. Harvey, Theresa B. Jain, and Jonalea R. Tonn, *The Effects of Thinning and Similar Stand Treatments on Fire Behavior in Western Forests*, Gen. Tech. Rept. PNW-GTR-463 (Portland, OR: USDA Forest Service, Pacific Northwest Research Station, 1999).

[44] Graham et al., *The Effects of Thinning on Fire Behavior.*

[45] For more information, see David M. Smith, Bruce C. Larson, Matthew J. Kelty, and P. Mark S. Ashton, *The Practice of Silviculture: Applied Forest Ecology*, 9th ed. (New York, NY: John Wiley & Sons, 1997).

[46] Russell T. Graham, Sarah McCaffery, and Theresa N. Jain, tech. eds., *Science Basis for Changing Forest Structure to Modify Wildfire Behavior and Severity*, Gen. Tech. Rept. RMRS-GTR-120 (Ft. Collins, CO: USDA Forest Service, Rocky Mountain Research Station, Apr. 2004), p. 25.

[47] Graham, et al., *Science Basis for Changing Forest Structure*, pp. 25, 27.

[48] Graham, et al., *The Effects of Thinning on Fire Behavior*, abstract.

[49] *A Strategic Assessment of Forest Biomass and Fuel Reduction Treatment in Western States*, Gen. Tech. Rept. RMRS-GTR-149 (Ft. Collins, CO: USDA Forest Service, Rocky Mountain Research Station, 2005).

[50] CRS Report RL32485, *Below-Cost Timber Sales: An Overview*, by Ross W. Gorte.

[51] Henry Carey and Martha Schumann, *Modifying WildFire Behavior — The Effectiveness of Fuel Treatments: The Status of Our Knowledge*, Southwest Region Working Paper 2 (Santa Fe, NM: National Community Forestry Center, April 2003), pp. I-ii.

[52] C. Larry Mason, Bruce R. Lippke, Kevin W. Zobrist, Thomas D. Bloxton, Jr., Kevin R. Cedar, Jeffrey M. Comnick, James B. McCarter, and Heather K. Rogers, "Investments in Fuel Removal to Avoid Forest Fires Result in Substantial Benefits," *Journal of Forestry*, v. 104 (Jan./Feb. 2006): p. 27.

[53] Carey and Schumann, *Modifying WildFire Behavior*, pp. I-ii.

[54] *A Strategic Assessment of Forest Biomass.*

[55] Roger D. Fight and R. James Barbour, *Financial Analysis of Fuel Treatments*, Gen. Tech. Rept. PNW-GTR-662 (Portland, OR: USDA Forest Service, Pacific Northwest Research Station, Dec. 2005).

[56] Data from annual agency budget justifications, presented in CRS Report RL33990, *Wildfire Funding*, by Ross W. Gorte, at pages CRS-6 and CRS-12.

[57] GAO, *Western National Forests: A Cohesive Strategy is Needed to Address Catastrophic Wildfire Threats*, GAO/RCED-99-65 (Washington, DC: April 1999). The Forest Service has done more fuel treatment in the South, where the generally gentler terrains, denser and more uniform timber stands, and historic fire patterns have kept treatment costs per acre lower than in the West.

[58] See CRS Report RL3 2436, *Public Participation in the Management of Forest Service and Bureau of Land Management Lands: Overview and Recent Changes*, by Pamela Baldwin.

[59] U.S. General Services Administration, Office of Governmentwide Policy, *Federal Real Property Profile, as of September 30, 2004*, pp. 18-19.

[60] USDA Natural Resources Conservation Service and the Iowa State University Statistical Laboratory, *Summary Report, 1997 Natural Resources Inventory (revised December 2000).*

[61] See CRS Memoranda, *Wildfire and Wildland Data*, by Ross W. Gorte, June 20, 2003, and March 3, 2008; available from the author.

[62] See CRS Report RL33990, *Wildfire Funding*, Table 7 (page CRS-12), by Ross W. Gorte.

[63] For information on ESA generally, see CRS Report RL3 1654, *The Endangered Species Act: A Primer*, by M. Lynne Corn, Eugene H. Buck, and Kristina Alexander.

[64] Amy Hessl and Susan Spackman, *Effects of Fire on Threatened and Endangered Plants: An Annotated Bibliography*, Information and Technical Report 2 (Ft. Collins, CO: U.S. Dept. of the Interior, National Biological Service, n.d.).

[65] For more information on NEPA generally, see CRS Report RL33 152, *The National Environmental Policy Act: Background and Implementation*, by Linda Luther, and CRS Report RS2062 1, *Overview of National Environmental Policy Act (NEPA) Requirements*, by Kristina Alexander.

[66] Executive Order 11514, "Protection and Enhancement of Environmental Quality," 35 *Fed. Reg.* 4247 (March 5, 1970).

[67] A simplified flowchart of this process can be found in Figure 1 (page CRS-22) of CRS Report RL33 152.

[68] 40 C.F.R. § 1508.4.

[69] 67 *Fed. Reg.* 54622 (Aug. 23, 2002).

[70] Forest Service Handbook, National Headquarters (WO), Washington, DC, *FSH 1909.15 - Environmental Policy and Procedures Handbook. Chapter 30 - Categorical Exclusion from Documentation*, Amendment No. 1909.15-2007-1 (Feb. 15, 2007).

[71] Sierra Club v. Bosworth, 510 F. 3d 1016 (9th Cir. 2007).

[72] For information on the status of relevant judicial decisions, see CRS Report RL33792, *Federal Lands Managed by the Bureau of Land Management (BLM) and the Forest Service (FS): Issues for the 110th Congress*, "FS NEPA Application and Categorical Exclusions," by Ross W. Gorte and Kristina Alexander.

[73] Section 322 of the Department of the Interior and Related Agencies Appropriations Act, 1993 (P.L. 102-381; 16 U.S.C. § 1612 note).

[74] GAO, *Forest Service: Information on Appeals and Litigation Involving Fuels Reduction Activities*, GAO-04-52 (Washington, DC: Oct. 2003).

[75] Gretchen M.R. Teich, Jacqueline Vaughn, and Hanna J. Cortner, "National Trends in the Use of Forest Service Administrative Appeals," *Journal of Forestry*, v. 102 (March 2004): pp. 14-19.

[76] David N. LaBand, Armando González-Cabán, and Anwar Hussain, "Factors That Influence Administrative Appeals of Proposed USDA Forest Service Fuels Reduction Actions," *Forest Science*, v. 52, no. 2 (2006): pp. 477-488.

[77] GAO, *Information on Appeals of Fuel Reduction Activities.*

[78] 68 *Fed. Reg.* 68254 (Dec. 8, 2003).

[79] See CRS Report RL33779, *The Endangered Species Act (ESA) in the 110th Congress: Conflicting Values and Difficult Choices*, by Eugene H. Buck, M. Lynne Corn, Pervaze A. Sheikh, Robert Meltz, and Kristina Alexander, at p. CRS-18.

[80] 40 C.F.R. § 1506.11.

[81] See CRS Report RL33 104, *NEPA and Hurricane Response, Recovery, and Rebuilding Efforts*, by Linda Luther.

[82] Email communication from Ted Bolling, CEQ, to Linda Luther, CRS, Jan. 22, 2008.

[83] See CRS Report 98-417 A, *Statutory Modifications of the Application of NEPA*, by Pamela Baldwin.

[84] David M. Ostergren, Kimberly A. Lowe, Jesse B. Abrams, and Elizabeth J. Ruther, "Public Perceptions of Forest Management in North Central Arizona: The Paradox of Demanding More Involvement but Allowing Limits to Legal Action," *Journal of Forestry*, v. 104 (Oct./Nov. 2006): pp. 375-382.

[85] Jeffrey J. Brooks, Alexander N. Bujak, Joseph G. Champ, and Daniel R. Williams, Collaborative Capacity, Problem Framing, and Mutual Trust in Addressing the Wildland Fire Social Problem: An Annotated Reading List, Gen. Tech. Rept. RMRS-GTR-182 (Ft. Collins, CO: USDA Forest Service, Rocky Mountain Research Station, Nov. 2006).

In: Wildfires and Wildfire Management
Editor: Kian V. Medina

ISBN: 978-1-60876-009-1

Chapter 5

WILDFIRE PROTECTION IN THE WILDLAND-URBAN INTERFACE

Ross W. Gorte

SUMMARY

Congress is giving increased attention and funding to wildfire threats. Much of the concern focuses on protecting homes and other structures in and near forests, an area known as the *wildland-urban interface*. However, not all agree on what can and should be done during wildfires, in their aftermath, and especially beforehand to protect the interface. This chapter describes the growth of the wildland-urban interface, wildfire suppression efforts, post-fire responses, and especially the programs and options for protecting the interface before the next wildfire strikes.

Wildfires have made national headlines in recent years, with major fires in the West and South killing firefighters, burning homes, and threatening communities. Federal funding for fire protection has more than doubled in the past decade, and administration and congressional leaders have urged additional wildfire protection. (See CRS Report RL33990, *Wildfire Funding*, by Ross W. Gorte.) Attention has focused on protecting people, homes, and communities in the *wildland-urban interface* (WUI), but opinions vary over how to protect the interface.

WHAT IS THE WILDLAND-URBAN INTERFACE?

The term *wildland-urban interface* (WUI) has been used for more than two decades to suggest an area where homes are in or near wildlands (forests or rangelands). The report from a 1986 conference on fire protection defined the WUI as "where combustible homes meet combustible vegetation."[1] In January 2001, the Forest Service (FS) and the Department of the Interior (DOI) identified types of interface communities.[2] Based on state data, they listed nearly 4,500 interface communities (with 11 states not providing data). In particular, the agencies defined an *interface community* as where wildlands abut structures with a clear line of demarcation between houses and wildland fuels, while an *intermix community* is where houses are scattered and intermingled with wildlands and fuels.

Recent research has found that the area of intermix communities is large and is growing faster than the area of interface communities.[3] In 2000, intermix communities in the three Pacific Coast states totaled 9.8 million acres, almost three times the 3.3 million acres in interface communities in those states. The 10-year growth in area of intermix communities was 14.1%, compared to only 2.5% for interface communities. However, the study acknowledged that determining the area of WUI communities was imprecise: "Mapping [the *Federal Register*] definition of the WUI using data and operational definitions we developed, we arrived at one possible representation of the WUI."[4] The intermingled nature of intermix communities poses significant challenges for fire protection efforts.

FIRE SUPPRESSION

In most of the Unites States, wildfires are inevitable. Biomass plus dry conditions equals fuel to burn. Add an ignition source (e.g., lightning or a thrown cigarette) and a wildfire happens. Fire is a self-sustaining chemical reaction that perpetuates itself as long as all three elements of the fire triangle — fuel, heat, and oxygen — remain available. Fire control focuses on removing one of those elements.

There are two principal kinds of wildfire, although an individual wildfire may contain areas of both kinds.[5] A *surface fire* burns the needles or leaves, grass, and other small biomass within a foot or so of the ground and quickly moves on. Such fires are relatively easy to control by removing fuel with a *fireline,* essentially a dirt path wide enough to eliminate the continuous fuels needed to sustain the fire, or by cooling or smothering the flames with water or dirt.

A *crown fire* burns biomass at all levels, from the surface through the tops of the trees. Crown fires do not consume all the biomass; rather, a crown fire quickly burns the needles or leaves and small twigs and limbs on the surface and throughout the crown of the trees. Because the needles and leaves in the crown are green, they require more energy to burn than dry fuels on the surface. Furthermore, because of the green fuels and the often discontinuous biomass of the canopy, wind is usually needed to sustain a crown fire. Once burning vigorously, a crown fire can create its own wind — the strong upward convection of the heated air can draw in cooler air from surrounding areas, thus creating a wind that feeds the fire. The strong upward convection can also lift burning biomass (*firebrands*) and send it soaring ahead of the fire, creating spot fires and accelerating the spread of the wildfire. Thus, crown fires are difficult, if not impossible, to control. Firelines are often ineffective, especially if winds are causing spot fires. Water or fire retardant (*slurry*) dropped from helicopters or airplanes can sometimes knock a crown fire down (back to a surface fire) if the area burning and the winds are not too great. Often, however, crown fires burn until they run out of fuel or the weather changes (the wind dies or it rains or snows).

Fires burn structures in one of three ways: through direct contact with fire (the fire burning right up to the structure); through radiation (heating from exposure to flames); and through firebrands landing on a flammable roof.[6] Surface fires generally only burn houses through direct contact, and protection is a relatively simple matter of a break in the continuous burnable material. In observing houses that burned in Los Alamos in 2000, one researcher stated "in several cases, a scratch line that removed [pine] needles from the base of a wood wall kept the house from igniting."[7] Crown fires, however, can burn houses in any of the three ways. The opportunity and ability to prevent structures from burning during a crown fire is small. Occasionally, water or some other wetting agent sprayed on walls or roofs can prevent ignition or extinguish firebrands from an advancing wildfire, but the firefighters could die of heat exposure or smoke inhalation from the approaching fire.

In the Aftermath

Recovery and efforts to support recovery after a severe wildfire vary, depending on the nature of the damages. For burned structures, insurance payment is the standard means for homeowners to pay for recovery — repair, if that is possible, or replacement, depending on the insurance policy. In a severe event, a presidential declaration of an emergency (in response to a request from a governor) initiates a process for federal assistance to state and local governments and to families and individuals to help with recovery. The nature and extent of the assistance depends on several factors, such as the nature and severity of damages and the insurance coverage of the affected parties.

For burned areas, site rehabilitation is sometimes warranted. In many temperate ecosystems, wildfires (including crown fires) are natural events, and the ecosystems are adapted to recover from the fire. Often, in severely burned areas, grass seed is spread to try to accelerate growth of ground cover and slow erosion, but grass often inhibits tree seed germination and growth, and thus may slow forest recovery. Rehabilitation efforts commonly focus on the firelines created to try to control the fire, since firelines are exposed bare earth that often run uphill, and thus can readily erode into gullies if left untreated. Some severely burned areas, particularly in coastal southern California, are susceptible to landslides during the subsequent rainy season. Monitoring can provide a warning to homeowners to evacuate an area prior to a landslide, but little can be done to prevent landslides in such situations.

Minimizing Wildfire Damages

Various efforts can protect structures and wildlands from some of the damages of wildfires. (See CRS Report RL34517, *Wildfire Damages to Homes and Resources: Understanding Causes and Reducing Losses*, by Ross W. Gorte.)

Protecting Structures

A structure's characteristics and landscaping significantly affect its chance of surviving a wildfire. Evidence from models, experiments, and case studies demonstrates that structural characteristics, especially the roofing materials, largely determine whether a home burns in a wildfire. Homes of brick or adobe with nonflammable roofs (e.g., tile, slate, metal) are far less likely to burn than homes with wood siding and flammable roofs (e.g., wood shingles).[8] Burnable materials (such as trees, shrubs, grass, pine needles, woodpiles, wood decks, and wooden deck furniture) within 40 meters (131 feet) of the structure also strongly influence whether the structure burns in a wildfire.[9]

Furthermore, the structure and landscape characteristics are more important than the intensity of the fire in determining whether a house burns. The Hayman Fire, in Colorado in June 2002, burned 132 houses — 70 houses (53%) were surrounded by crown fire, while 62 houses (47%) were surrounded by surface fire.[10] In addition, 662 homes (83% of all homes within the fire perimeter) survived the fire, even though 35% of the area was severely burned and 16% was moderately burned.[11] This suggests that at least some of the structures survived despite a crown fire around them; why these structures survived was not reported.

Protecting Wildlands

The impact of wildfires on wildlands depends largely on the nature of the ecosystem. Some ecosystems are adapted to and recover from periodic crown fires — perennial grasslands, chaparral, lodgepole and jack pines, and more. In these ecosystems, the plants have evolved to resprout or reseed the burned areas, and thus recover from crown fires by outcompeting other plant species. Eliminating crown fires could eventually eliminate these ecosystems. However, eliminating crown fires in these ecosystems is probably impossible, since the plants contribute to the development and spread of crown fires — grasses burn quickly; chaparral has a high volatile-oils content; and lodgepole and jack pines grow in dense, even-aged stands.

Other ecosystems are adapted to relatively frequent (5- to 35-year intervals) surface fires. Fire suppression has been moderately successful in controlling surface fires, and thus the needles, twigs, and other fine and small fuels have been accumulating for three or more fire cycles. This abnormal fuel accumulation, combined with fuel ladders of brush, small trees, and low limbs (many of which would have burned in a surface fire), have led to crown fires where such fires were historically rare. Fuel reduction treatments can restore conditions in frequent-surface-fire ecosystems to again make crown fires rare occurrences, reducing damages to resources.

Protecting the WUI

Reducing fuels in the WUI has been a controversial aspect of congressional debates over fire protection legislation. The evidence discussed above indicates that fuel reduction provides little protection for structures. However, some observers have noted that the WUI is more than just a collection of houses:[12]

> A town is not just the place where people have homes. Communities are in the forest because they are emotionally, economically, and socially linked and dependent on the forest. When we consider the areas that need immediate treatment we should consider the human community "impact area" — the entire area that, if impacted by a catastrophic fire, will undermine the health and livelihood of a community.

At a minimum, most would agree on the need for an area of *defensible space* around homes that needs to be cleared of burnable materials — at least 10 meters (33 feet) and possibly as much as 40 meters (131 feet). One observer recommended that protecting communities should include intensive treatment to reduce fuels and burnable materials in the *home ignition zone*, up to 200 meters (655 feet) around structures, with less intensive fuel treatment in the *community protection zone*, generally up to 500 meters (1,640 feet, or about a third of a mile) from structures.[13]

The Healthy Forests Restoration Act of 2003 (HFRA; P.L. 108-148; 16 U.S.C. § 6511) established a somewhat broader standard for fuel reduction activities under its authorities. Section 101(16) of HFRA defined the WUI to include an area out to 1/2 mile from the boundary of an *at-risk community* or 1 1/2 miles from the boundary if a sustained steep slope could cause dangerous fire behavior or to an effective fire break, such as a road or ridge top. HFRA included no guidance on how to apply these standards in intermix communities, with no definitive boundary.

Issues for Congress

As more acres and homes have burned in recent years, and more people are at risk from wildfires, Congress is facing increasing pressures for wildfire protection. Congress decides what programs to authorize and fund. Many programs exist, and other options are possible.

Firefighting uses the majority of wildfire management funding, accounting for $1.1 to $1.9 billion annually (including emergency supplemental funds) since FY2003. Appropriations for fire suppression have risen in nearly every year for a decade, going from $277 million in FY1999 to the requested $1.33 billion for FY2009. Given the difficulty in suppressing crown fires, one might question the effectiveness of continued increases in suppression funding, although the agencies also clearly need to show the public that they are doing all they can to stop the threatening and damaging fires.

Federal programs to protect homes are currently limited to information, primarily through FIREWISE, for homeowners on how to protect their homes. Programs could be expanded to educate homeowners, state and local governments, and the insurance industry about the ways to protect homes through actions, planning, and zoning and building regulations. Congress could create and fund new programs to assist homeowners in renovations to make their homes fire-safe and to create defensible space around their structures, through direct federal assistance or through the states.

Congress could also consider expanding protection for defensible space beyond the home ignition zone to a community protection zone. HFRA authorizes an expedited review process for activities on federal lands in the WUI. Perhaps other changes could further accelerate action. Funding for fuel reduction in the WUI could also be expanded. Appropriations for fuel reduction have averaged$500 million annually since FY2006, but only a portion is used in the WUI, and funding is far below the estimated amount needed to treat the lands at risk. (See the discussion in CRS Report RL33990, *Wildfire Funding*, by Ross W. Gorte). State fire assistance funding through the Forest Service could be used for fuel reduction in the WUI, at the discretion of the states, but funding has averaged $88 million annually and the states have many wildfire priorities. Additional funding through the states for fuel reduction on private lands in the WUI is a possibility that Congress could contemplate.

In addition, Congress might debate choices for compensating homeowners for property losses due to wildfires. One option might be to restrict compensation to those who had acted to protect their homes, but got burned anyway. Another option might be to require that compensation for rebuilding be used only for fire-safe building designs and materials. Alternatively, Congress could establish a national wildfire insurance program, with premiums based on fire threats, the fire-safety of the structures, and the defensible space being maintained.

Finally, Congress could consider compensation for landowners that suffer resource losses from wildfires. An emergency reforestation assistance program has existed for many years, although it has not been funded since FY1 993. (See CRS Report RL3 1065, *Forestry Assistance Programs*, by Ross W. Gorte.) In the 2008 farm bill, Congress included forest restoration assistance in an existing emergency conservation program. (See CRS Report RL33917, *Forestry in the 2008 Farm Bill*, by Ross W. Gorte.) These programs can provide assistance in recovery from a wildfire disaster, but do not compensate landowners for losses in the way that homeowners are compensated for the loss of their homes. Congress might consider such additional compensation.

End Notes

[1] USDA Forest Service, National Fire Protection Association, and U.S. Fire Administration, *Wildfire Strikes Home!* (Jan. 1987), p. 2.

[2] U.S. Dept. of Agriculture and Dept. of the Interior, "Urban Wildland Interface Communities Within the Vicinity of Federal Lands That Are at High Risk From Wildfire," 66 *Fed. Reg.* 753 (Jan. 4, 2001).

[3] Roger B. Hammer, Volker C. Radeloff, Jeremy S. Fried, and Susan L. Stewart, "WildlandUrban Interface Housing Growth During the 1990s in California, Oregon, and Washington," *International Journal of Wildland Fire*, v. 16 (2007): pp. 255-265.

[4] Hammer et al., "Wildland-Urban Interface Housing Growth," p. 256.

[5] See Stephen F. Arno and Steven Allison-Bunnell, *Flames in Our Forest: Disaster or Renewal?* (Washington, DC: Island Press, 2002), pp. 45-46.

[6] National Wildland/Urban Interface, Fire Protection Program, *Wildland/Urban Interface Fire Hazard Assessment Methodology*, p. 5, at [http://www.firewise.org/resources/files/wham.pdf].

[7] Jack Cohen, "The Cerro Grande Fire: Why Houses Burned," *Forest Trust Quarterly Report*, no. 13 (Dec. 2000): p. 11.

[8] Jack D. Cohen, "Preventing Disaster: Home Ignitability in the Wildland-Urban Interface," *Journal of Forestry*, v. 98, no. 3 (Mar. 2000): 15-21.

[9] Cohen, "Preventing Disaster."

[10] Jack Cohen and Rick Stratton, "Home Destruction Within the Hayman Fire Perimeter," *Hayman Fire Case Study*, Gen. Tech. Rept. RMRS-GTR-1 14 (Ft. Collins, CO: USDA Forest Service, Sept. 2003), p. 264.

[11] Peter Robichaud, Lee MacDonald, Jeff Freeouf, Dan Neary, Deborah Martin, and Louise Ashman, "Postfire Rehabilitation of the Hayman Fire," *Hayman Fire Case Study*, Gen. Tech. Rept. RMRS-GTR-1 14 (Ft. Collins, CO: USDA Forest Service, Sept. 2003), p. 294.

[12] W. Wallace Covington, Director, The Ecological Restoration Institute, Northern Arizona University, "Prepared Statement," *National Fire Plan*, hearing before the Senate Committee on Energy and Natural Resources, July 16, 2002, S.Hrg. 107-834 (Washington, DC: GPO, 2003), p. 61.

[13] Brian Nowicki, *Effectively Treating the Wildland-Urban Interface to Protect Houses and Communities from the Threat of Forest Fire* (Tucson, AZ: Center for Biological Diversity, Aug. 2002).

In: Wildfires and Wildfire Management
Editor: Kian V. Medina
ISBN: 978-1-60876-009-1

Chapter 6

Wildland Fire Management: Interagency Budget Tool Needs Further Development to Fully Meet Key Objectives

United States Government Accountability Office

Why GAO Did This Study

Wildland fires have become increasingly damaging and costly. To deal with fire's threats, the five federal wildland fire agencies—the Forest Service in the Department of Agriculture and four agencies in the Department of the Interior (Interior)—rely on thousands of firefighters, fire engines, and other assets. To ensure acquisition of the best mix of these assets, the agencies in 2002 began developing a new interagency budget tool known as fire program analysis (FPA). FPA underwent major changes in 2006, raising questions about its ability to meet its original objectives. GAO was asked to examine (1) FPA's development to date, including the 2006 changes, and (2) the extent to which FPA will meet its objectives. To do so, GAO reviewed agency policies and FPA documentation and interviewed agency officials.

What GAO Recommends

GAO is recommending, among other things, that the agencies develop a strategic plan for the continued development of FPA and provide Congress with annual updates on (1) their progress in completing the steps outlined in that plan and (2) how they used FPA in developing their budgets. Interior disagreed with the need to develop a strategic plan. In response to Forest Service and Interior comments on GAO findings on FPA's cost-effectiveness approach, GAO's recommendation to develop a strategic plan was revised to provide more flexibility. The agencies generally concurred with the other recommendations.

WHAT GAO FOUND

FPA is both a computer model and a broader management system for developing the five agencies' wildland fire budget requests and allocating funds. FPA is intended to allow the agencies to analyze potential combinations of firefighting assets and potential strategies for reducing vegetation and fighting fires to determine the most cost-effective mix of assets and strategies. The agencies began developing FPA in 2002 and completed the first part of the model in October 2004. As the agencies began using FPA, however, agency officials raised concerns about its underlying science and the extent to which it met agency management and policy objectives. As a result, in 2006 the agencies conducted a review of FPA, which questioned FPA's basic modeling approach. The agencies made substantial changes to FPA after the review, some of which followed from the review's recommendations. For example, as recommended, the agencies established a new oversight body comprising senior agency leaders. The agencies also made fundamental changes to FPA's modeling approach for analyzing the firefighting assets needed to respond to fires, but these changes went beyond the review's recommendations and, despite FPA's importance and cost, the reasons for these changes were not fully documented. The agencies expected to complete the FPA model in November 2008—about a year later than initially estimated—and to begin using FPA's results in spring 2009 to develop their fiscal year 2011 budget requests, a delay of about 3 years from their initial goal of using FPA's preliminary results in 2006. FPA is expected to cost about $54 million to develop.

Although it is not yet complete and GAO conducted only a limited review of its available components, FPA shows promise in achieving some of the key objectives originally established for it; nevertheless, the approach the agencies have taken hampers FPA from meeting other key objectives. Among the most important objectives, FPA will (1) provide a common framework for the five federal agencies to analyze firefighting assets and develop budget requests across agency jurisdictions, (2) analyze the most important fire management activities, and (3) recognize the presence of certain nonfederal firefighting assets that may be available to respond to fires on federal land. FPA falls short, however, with respect to other key objectives. First, FPA has limited ability to project the effects of different levels of vegetation reduction treatments and firefighting strategies over time, meaning that agency officials lack information that could help them analyze the long-term impact of changes in their approach to wildland fire management. Second, the modeling approach the agencies are taking cannot identify the most cost-effective mix and location of federal firefighting assets for a given budget but, rather, analyzes a limited number of combinations of assets and strategies to identify the most cost-effective among them. More broadly, the current FPA approach involves considerable discretion on the part of agency officials, increasing the importance of making decisions in a transparent manner so that Congress, the public, and officials throughout the agencies understand FPA's role in budget development and allocation.

ABBREVIATIONS

Agriculture	Department of Agriculture

FPA	fire program analysis
Interior	Department of the Interior
OMB	Office of Management and Budget

November 24, 2008

The Honorable Jeff Bingaman
Chairman
Committee on Energy and Natural Resources
United States Senate

Dear Mr. Chairman:

Wildland fires increasingly threaten communities and natural resources, and the cost of responding to those fires has risen dramatically. To deal with fire's threats, the five federal agencies responsible for managing wildland fires—the Forest Service in the Department of Agriculture (Agriculture) and the Bureau of Indian Affairs, Bureau of Land Management, Fish and Wildlife Service, and National Park Service in the Department of the Interior (Interior)—call upon thousands of firefighters and station fire engines, aircraft, and other equipment on or near federal land across the country. The agencies also conduct treatments to reduce vegetation, in an effort to lessen the potential for severe wildland fires, decrease the damage caused by fires, and restore and maintain healthy ecosystems. Despite these efforts, the average number of acres burned annually in recent years has grown by about 70 percent, and federal appropriations to prepare for and respond to wildland fires have nearly tripled since the mid-1990s, to more than $3 billion annually. Several factors have contributed to the increased risk and cost, including uncharacteristic accumulations of vegetation, in part due to past land management activities and fire suppression policies; (2) increasing human development in or near wildlands, an area commonly known as the wildland-urban interface; and (3) severe drought and other stresses, in part due to climate change. Combined, these factors have contributed to wildland fires' burning more intensely and spreading more quickly at the same time that development has continued in fire-prone areas. Longstanding concerns about the mounting risk from and cost of wildland fires, along with growing recognition of the long-term fiscal challenges facing the nation, have led Congress, the agencies, and others to focus on ensuring that federal wildland fire activities are appropriate and carried out in a cost-effective and efficient manner.

A key initial step in this effort was the development of the 1995 federal wildland fire management policy.[1] The policy recognized that new approaches to managing wildland fire were needed if the agencies were to respond effectively to changing conditions. The policy also found that differences in budgeting processes among the five agencies hindered their response to wildland fires. Subsequently, congressional committees directed the agencies to develop a common budget process. In 2001, the agencies commissioned a report that established a vision for an interagency budget process, a report the agencies adopted as the basis for a new budget-planning system known as fire program analysis, or FPA.[2]

As envisioned in the 2001 agency report, as well as in congressional committee and Office of Management and Budget (OMB) reports, FPA was intended to help the agencies develop their wildland fire budget requests and allocate funds. FPA's objectives include

- providing a common budget framework to analyze firefighting assets without regard for agency jurisdictions;
- examining the full scope of fire management activities, including preparing for fires by acquiring and positioning firefighting assets for the fire season, mobilizing assets to suppress fires, and reducing potentially hazardous fuels;
- considering the availability of nonfederal firefighting assets, such as state or county firefighters, that typically help respond to fires on federal lands;
- considering the communities and resources to be protected and agency land management objectives;
- modeling the effects over time of differing strategies for responding to wildland fires and treating lands to reduce hazardous fuels; and
- using this information to identify the most cost-effective mix and location of federal wildland fire management assets.

In addition, FPA was expected to be externally peer reviewed, which could improve Congress's and the agencies' understanding of its strengths and weaknesses.

To realize this vision, the agencies in 2002 began to develop FPA, designing it as a computer model that analyzed numerous potential combinations and locations of firefighting assets and, for any given budget level, identified the optimal mix of these assets—that is, the mix and locations of firefighting assets that would best protect resources at risk. Data on potential combinations and locations of assets were to be entered by fire officials at agency field units, and the model's analysis of these combinations would then be evaluated by agency budget officials at the national level. The agencies estimated that FPA would cost more than $40 million to develop and would take about 5 years to complete. In 2006, after 4 years of work, the agencies conducted an internal review of FPA, in part because of concerns about how well the computer model reflected the realities of the agencies' fire management activities.[3] Subsequently, the agencies made substantial changes in how FPA analyzes needed firefighting assets and determines where best to locate them. These changes raised questions about the extent to which FPA would meet its original objectives. In this context, you asked us to review FPA. This chapter examines (1) how the agencies have developed FPA to date, including the process followed as part of the internal review, and FPA's current status; and (2) the extent to which FPA will meet its original objectives.

To address our objectives, we reviewed agency documents on FPA development, including the interagency report and project charter that provide FPA's foundation, the reports resulting from the internal review, and numerous technical papers and other documentation describing particular aspects of FPA. To further our understanding of FPA's development, including changes made to FPA after the internal review, we interviewed Forest Service and Interior officials in Washington, D.C.; FPA project staff in Boise, Idaho; and agency officials in the field who were familiar with FPA. We also interviewed agency and other scientists who have helped develop FPA. At the time of our review, however, substantial portions of the model remained incomplete, and the agencies had not documented the model sufficiently to allow a comprehensive evaluation. We therefore limited our review to a broad examination of FPA's various components and how they interact, as well as a comparison of FPA's current approach and capabilities with its original objectives. We did not compare the capabilities of the current approach to those of the approach taken before the internal review. Appendix I describes our scope and methodology in more detail. We conducted this performance audit from September 2007 through November 2008 in accordance with generally accepted government auditing standards. Those standards require that we plan and perform the audit to obtain sufficient, appropriate evidence to provide a reasonable basis for our findings and conclusions based on our audit objectives. We believe that the evidence obtained provides a reasonable basis for our findings and conclusions based on our audit objectives.

RESULTS IN BRIEF

The Forest Service's and Interior agencies' initial development and implementation of FPA gave rise to concerns about its performance, leading to an internal review and subsequent changes to the model. These changes, however, went beyond the review's recommendations and were not always clearly explained or fully documented. The agencies began developing FPA in 2002 and completed the first part of the model in October 2004. But as field units began to use the first part, senior agency officials and some field staff raised fundamental concerns—including concerns about the underlying science and the extent to which FPA met agency management and policy objectives. As a result, in 2006 the agencies conducted an internal review of FPA, which questioned its modeling approach and concluded, among other things, that agency leadership needed to become more involved if FPA were to succeed. The agencies made substantial changes to FPA after the review, some of which followed from the review's recommendations. For example, the agencies established a new oversight body comprising senior agency leaders and an interagency science team. The agencies, with the approval of the oversight body, also made fundamental changes to FPA's modeling approach for analyzing the firefighting assets needed to respond to fires, but these changes went beyond the review's recommendations. The review, for example, did not conclude that a different modeling approach was needed, instead recommending that the agencies continue testing the initial model and refine it as necessary. Rather than follow this recommendation, however, the agencies adopted an entirely new modeling approach. Yet despite FPA's importance and cost, the reasons for these changes were not fully documented, and a formal, documented comparison of the original and revised approaches was never conducted. The agencies expected to complete the FPA model in November 2008—about a year later than initially estimated—and to begin using FPA's results in spring 2009 to develop their fiscal year 2011 budget requests, a delay of about 3 years from their initial goal of using FPA's preliminary results in 2006. Ultimately, FPA is expected to cost about $54 million to develop.

Although it is not yet complete and we conducted only a limited review of its available components, FPA shows promise in achieving some of the key objectives that congressional committees, OMB, and the agencies themselves established for it. Nevertheless, the approach the agencies have taken hampers FPA from meeting other key objectives. Once FPA is more fully developed and documented, a detailed, external peer review may reveal more about the extent to which it will help the agencies develop their wildland fire budget requests and allocate funds. Among the most important objectives it is likely to achieve, FPA is to provide a common framework for the five federal agencies to develop their wildland fire budget requests and analyze needed firefighting assets across agency jurisdictions—a significant step forward—and is to analyze the three most important fire management activities (preparedness, fire suppression, and fuel reduction). The agencies also have developed FPA to be capable of recognizing the presence of nonfederal firefighting assets that may be available to respond to fires on federal land—another key objective— although the extent to which these assets is to be included in the analysis is not yet clear. And finally, FPA is also to consider specific land management objectives and resources at risk, as suggested by the 1995 federal wildland fire management policy, rather than simply assume that all fires should be suppressed as quickly as possible (although if implemented as currently developed, FPA will likely not allow the agencies to consistently identify the locations that are most important to protect from a national perspective). FPA falls short, however, with respect to other key objectives in two critical areas. First, FPA's ability to project the effects of different levels of fuel reduction treatments and firefighting strategies over time appears limited. Agency officials are therefore likely to lack information that would help them analyze the extent to which increasing or decreasing funding for fuel reduction treatments and responding more or less aggressively to fires in the short term could affect the expected cost of responding to wildland fires over the long term. Second, regardless of the extent to which other key objectives are met, the modeling approach the agencies have taken is unlikely to identify the most cost-effective mix and location of federal firefighting assets for a given budget but only whether a particular mix of assets is more or less cost-effective than another. Since the different mixes of assets analyzed are limited to the number of alternatives developed by agency units in the field, these alternatives, even taken together, are unlikely to include the single most cost-effective mix of assets nationwide. In addition, other aspects of FPA may complicate its further development and implementation, including the lack of an external peer review of the model to date. Agency officials recognize many of these shortcomings and have said that they are considering taking actions—such as further adjusting the model (to better identify the most highly valued resources to protect, for example) and submitting the model for peer review—that have the potential to move FPA closer to meeting its key objectives. Regardless of the specific objectives FPA achieves, the modeling approach the agencies selected for FPA involves considerable discretion on the part of agency decision makers, increasing the importance of making decisions in a manner transparent enough that Congress, the public, and officials throughout the agencies understand how the decisions were made and FPA's role in them.

To improve the agencies' ability to use FPA in developing their wildland fire management budget requests and allocating funds in a cost-effective manner and to promote transparency in decision making—and recognizing that FPA is still under development and that completing it will be an iterative process requiring the agencies' continued effort to improve—we are recommending that the Secretaries of Agriculture and the Interior (1) direct the agencies to develop a strategic plan for the continued development of FPA, (2) report annually to Congress on their progress in completing the steps outlined in this plan and on FPA's ability to meet each of its key original objectives, (3) report to Congress each year on how the agencies used FPA to develop their budget requests and allocate funds, and (4) submit the model for external peer review.

In written comments on a draft of this chapter, the Forest Service and Interior disagreed with our finding that FPA is unlikely to allow the agencies to identify the most cost-effective mix of firefighting assets, stating they believed that FPA will allow them to meet the goal of cost- effectiveness. They also commented that the revised approach they are taking in developing FPA is more realistic and appropriate than their original approach. We continue to believe, however, that, regardless of the comparative strengths and weaknesses of the original and revised approaches, FPA as it is being developed is unlikely to allow the agencies to identify the most cost-effective location and mix of assets and strategies—one of the agencies' original objectives for FPA. To account for the agencies' views that the revised approach is more realistic, we are modifying our recommendation that the agencies develop a strategic plan for the continued development of FPA, adding that the agencies should clearly state whether they believe any of FPA's key original objectives are no longer appropriate. The Forest Service commented that it fundamentally agreed with our recommendations but believes there are better alternative approaches to carrying some of them out. The agency described the steps it intended to take in addressing two of them, but we do not believe that the steps outlined in the letter are specific and transparent enough to meet the intent of our recommendations. Interior concurred with three of our recommendations but disagreed with our recommendation that the agencies develop a strategic plan for the continued development of FPA, stating that developing such a plan would delay deployment and increase the cost of FPA. We do not agree that creating a strategic plan would necessarily delay the agencies' implementation of FPA; further, because our review raised questions about FPA's ability to meet certain key objectives, we continue to believe it is important for the agencies to create a strategic plan that directly and transparently evaluates FPA's ability to meet its original objectives and identifies ways to improve FPA to better meet those objectives. Comments from the Forest Service and Interior, along with our responses to those comments, are reprinted in appendixes II and III, respectively.

BACKGROUND

The agencies' wildland fire management program has three major components: preparedness, suppression, and fuel reduction.[4] To prepare for a wildland fire season, the agencies acquire firefighting assets— including firefighters, engines, aircraft, and other equipment—and station them either at individual land management units (such as national forests or national parks) or at centralized dispatch locations. The primary purpose of these assets is to respond to fires before they become large—a response referred to as initial attack—thus forestalling threats to communities and natural and cultural resources. The speed with which the agencies are able to respond to a fire can be critical to their ability to suppress it while it is small; increasing the number of firefighting assets available to respond, and the number of locations they can respond from, can therefore improve the agencies' initial attack success, although the marginal utility of adding more firefighting assets decreases as the number of assets goes up. The assets the agencies use for initial attack are funded primarily from the agencies' preparedness budget accounts.

In the relatively rare instances in which fires escape initial attack and grow large, the agencies respond using an interagency system, in which additional firefighting assets from federal, state, and local agencies, as well as private contractors, are mobilized, regardless of which agency or agencies have jurisdiction over the burning lands.[5] Federal agencies typically fund the costs of these activities from their suppression budget accounts. To reduce the potential for severe wildland fires, lessen the damage caused by fires, limit the spread of flammable invasive species, and restore and maintain healthy ecosystems, the agencies also reduce potentially hazardous vegetation that can fuel fires. They remove or modify fuels using prescribed fire, mechanical thinning, herbicides, certain grazing methods, or combinations of these or other approaches.

The federal government's cost of preparing for and responding to wildland fires has increased substantially over the past decade—an increase that has led federal agencies to fundamentally reexamine their approach to wildland fire management. For decades, federal agencies aggressively suppressed wildland fires and generally succeeded in reducing the number of acres burned. In some parts of the country, however, rather than eliminating severe wildland fires, decades of suppression contributed to the disruption of ecological cycles and began to change the structure and composition of forests and rangelands, thereby making lands more susceptible to fire. Increasingly, the agencies have recognized the role that fire plays in many ecosystems and the role that it could play in the agencies' management of forests and watersheds. As a result, the agencies have increased their efforts to reduce fuels and their emphasis on using less aggressive firefighting strategies, which typically cost less and can reduce fuels across a broader area than if fires are aggressively suppressed. Such strategies are to be used only in appropriate situations, such as in responding to fires that are not expected to threaten communities or damage important natural or cultural resources.

This approach to managing wildland fires requires close integration of planning and budgeting systems so that the agencies are able to holistically analyze the full wildland fire management program. The agencies historically have used different planning and budgeting systems to help develop their budget requests and allocate the funds Congress appropriates. The agencies have identified shortcomings with this approach and have recognized that the existing systems were not capable of analyzing the trade-offs among initial attack, the full range of suppression strategies, and fuel reduction. Aggressively suppressing a fire, for example, may cost less in the short term but contribute to continued accumulation of vegetation, which can increase both the risk from and cost of responding to fires in the future; conversely, increasing investment in reducing fuels may cost more in the short term but can provide future benefits. The agencies, following congressional committee direction, committed to developing a new system, which came to be known as FPA. FPA is a strategic tool that agency budget officials expect to use to develop their wildland fire budget requests and allocate their fire management funds to the field, and that agency fire officials expect to use to model the effect that differing mixes and locations of firefighting assets, and differing levels of investment in reducing fuels, will have on their ability to protect communities and resources. Because it is a strategic tool rather than a tactical one, agency fire managers would not use FPA to help the agencies respond to actual fires.

In developing and using FPA, the agencies must consider the process and time frames of the annual federal budget cycle, which begins about 2 years before the fiscal year for which funds are being requested. Agencies develop their budget requests in late spring and summer and submit them to OMB in September. OMB prepares budget materials to submit to the President in January. The President approves a budget proposal and sends it to Congress by the first Monday in February. To develop their fiscal year 2011 budgets, for example, the agencies, in conjunction with their respective departments, expect to begin developing their budget requests in spring 2009 and to submit them to OMB in September 2009; subsequently, the President would submit his budget request to Congress in February 2010 for Congress's consideration.

CONCERNS ABOUT FPA'S EARLY PERFORMANCE LED TO SIGNIFICANT CHANGES, NOT ALL OF WHICH WERE TRANSPARENT

Concerns about FPA's early performance and about the policy and scientific approaches the agencies used in FPA's early development led the agencies to conduct an internal review of FPA in 2006. Subsequently, the agencies made several significant changes to FPA, but these changes went beyond those recommended by the review, and the reasons for several of the changes were not fully documented. The agencies do not expect to use preliminary FPA results to develop their budget requests until 2009— 3 years later than they had initially planned—but the cost of completing FPA appears in line with previous estimates.

Concerns during FPA's Initial Implementation Led to an Internal Review

The staff who began to develop FPA in 2002, following congressional committee direction, initially focused on developing the portion of the model that analyzed the agencies' ability to successfully contain wildland fires during initial attack. The staff selected an approach that relied primarily on a modeling technique known as optimization. Using this approach, FPA was to analyze, for any given budget level, all possible combinations and locations of the firefighting assets typically available to agency field units and identify the combination of these assets that resulted in optimal protection of communities and resources. To provide data on different potential firefighting assets and locations, the agencies divided the country into 139 interagency "fire planning units," each of which encompassed land managed by one or more of the federal agencies responsible for wildland fire.[6] Fire management officials in each of these planning units then identified the relative importance of protecting each acre within that planning unit by assigning a weighting factor indicating each acre's importance relative to other acres. The most important acres to protect, such as those in the wildland-urban interface, were assigned a weight of 1.0, while less important acres were assigned proportionately lower weights. After analyzing historic fire occurrence and weather patterns associated with each planning unit to determine where and when fires were likely to start, and considering the relative importance of acres to be protected, FPA was to analyze, for any given budget level, all possible mixes and locations of firefighting assets typically available to those units in order to determine which mix and locations would afford the best level of protection.

Development of this "preparedness module" was completed in October 2004, and over the next 16 months, officials in the field began using it to analyze their preparedness assets and budgets. By February 2006, nearly all the fire planning units had submitted FPA results for their units to the agencies' Washington offices, which in turn analyzed the FPA results in an effort to identify the optimal mix and location of firefighting assets across the country. During this time, however, senior agency officials, as well as some field officials, began to raise fundamental concerns about FPA's modeling approach. Weighting the importance of individual acres within a fire planning unit, for example, was a central component of FPA's early approach, and some officials believed that even where resources to be protected were similar across planning units, officials in those units assigned substantially different weights to the resources, thereby undermining the reliability of the results. Other officials were concerned that the early FPA approach placed insufficient emphasis on containing fires during initial attack, although the officials who developed this approach noted that it reflected the interagency policy of responding to fires on the basis of specific land and fire management objectives, rather than simply assuming that all fires should be suppressed as quickly as possible. Still other officials were concerned that the initial approach could result in unrealistic shifts in the mix and location of assets; a small change in budget, for example, could have led FPA to suggest moving a large quantity of assets from one planning unit to another or to dramatically change the relative proportion of firefighters, engines, and aircraft within a planning unit.

Despite these concerns, senior agency officials told us, the early development of FPA represented an important first effort, given the difficulty of the project; in hindsight, they also recognized that greater involvement by policy and budget officials and agency scientists might well have averted some of the concerns and helped FPA develop more quickly. In any case, despite having told congressional and OMB staff that they intended to use the results of this initial analysis to help develop their 2008 budget requests, agency officials decided that the concerns about FPA were too great to justify doing so, and they instead initiated a two-part review of FPA to evaluate the issues that had surfaced.

The agencies conducted this two-part review of FPA in late 2005 and early 2006. The reviews—performed by agency land managers, fire and budget officials, and scientists, as well as a representative from a state forestry agency—consisted of (1) an evaluation of the extent to which FPA helped the agencies achieve their management and policy objectives and (2) an evaluation of particular aspects of the underlying science and modeling approach. The reviews reaffirmed FPA's original objectives as articulated in the 2001 report, but they identified several challenges to meeting these objectives and made several recommendations intended to strengthen FPA's ability to do so. The management review, for example, recommended that the agencies more fully involve senior officials and scientists, submit FPA for external peer review,[7] and complete their analysis of the initial FPA results. The science review likewise recommended that FPA be peer-reviewed and, in addition, that the agencies further test and improve the model and the data it uses. The agencies conducted the reviews quickly, however, and did not intend them to be a comprehensive evaluation of FPA; the science review, in particular, examined only certain aspects of FPA.

The Agencies Made Significant Changes to FPA, Not All of Which Were Fully Documented

After the reviews, the agencies made several changes to the process used for developing FPA, changes that generally followed from the reviews' recommendations. In April 2006, the agencies established a new oversight body comprising senior officials from the Forest Service and Interior. This group was formed to make strategic decisions about FPA's scope, determine how FPA would be used to help the agencies make funding allocation decisions, and address any policy issues that FPA's development raised. The group also was to keep the Wildland Fire Leadership Council[8] informed of FPA's status, including issues that the council needed to resolve. The agencies also established an interagency science team, made up of scientists from both the Forest Service and Interior, as well as university scientists outside the agencies. This science team was to assist FPA's developers by reviewing and evaluating FPA's modeling approach and identifying data sources and analytical techniques that could further FPA's development.

The agencies also changed FPA's modeling approach considerably. Rather than continue to use the initial optimization-based approach (evaluating, for a given budget level, all possible combinations and locations of firefighting assets typically available to local units and identifying the asset combination that provided optimal protection of communities and resources), the agencies switched to a simulation modeling approach that evaluates a much smaller number of potential asset combinations along with different options for fuel reduction treatments and ranks them according to certain performance criteria—which also differ from those used previously. The new approach is no longer to simply assess the extent to which each asset combination protects the areas field officials have identified as most important. Instead, it is to evaluate each combination's predicted performance against five separate performance measures the agencies have established:

- total projected cost of suppressing fires;
- total number of acres burned in the wildland-urban interface;
- total number of acres meeting fire and fuels management objectives, such as reducing the likelihood of intense fires;
- total number of acres burned containing resources the agencies define as being highly valued, such as endangered species habitat or municipal watersheds; and percentage of fires contained while small (i.e., the initial attack success rate).[9]

The revised FPA approach encompasses both a computer model and a management system to help the agencies develop their budget requests and allocate funds. Agency officials in each of the 139 planning units are to develop, for each of a given number of budget levels, an option specifying the mix and location of firefighting assets they would choose to acquire and an option specifying the number of acres they would treat to reduce fuels. For example, officials might develop one mix and location of firefighting assets and the acreage that would be treated if their fire planning unit's budget remained the same as the previous year, another option corresponding to a budget decrease from the previous year, and a third option corresponding to a budget increase from the previous year. The number of options the planning units are to develop and the budget levels to which these options correspond will depend on annual field guidance prepared by the agencies' headquarters offices. Senior agency officials told us that during FPA's initial implementation they were considering directing the units to develop three preparedness options and three fuel treatment options. These options were to correspond to each unit's 2007 budget level and plus and minus 10 percent of these 2007 levels. As of November 2008, however, the agencies had not finalized this step.

Once the planning units have developed their options and entered information about firefighting assets and fuel treatments into the FPA system, the computer model is to then analyze historical data on local fire occurrence; local vegetation, geography, and weather; and the predicted effect on fire behavior of reducing fuels. From this analysis, FPA is to model the likelihood that wildland fire will damage communities and resources within the fire planning unit, considering the different mixes of assets and fuel treatments reflected in the proposed options. To provide comparable information across planning units, FPA is to evaluate each unit's options against the five performance measures.

The FPA model is to then calculate a performance score for each of the "alternatives" developed by each planning unit. (An alternative consists of one preparedness option paired with one fuel treatment option. If planning units were directed to prepare three preparedness and three fuel treatment options, for example, nine alternatives would be possible.) FPA is to then "roll up" the performance scores for each alternative in all 139 planning units, so that senior agency officials can evaluate the effects on the agencies' performance measures nationwide of different combinations of alternatives. The senior agency officials would then use FPA results in conjunction with other budget information and processes to develop their budget requests.

The extent to which any particular alternative, or set of alternatives, is considered cost-effective relies on the weights assigned to each of the five measures. The agencies could weight these measures in several ways to reflect their relative importance. If one measure were overwhelmingly more important than the others—if the agencies wanted to minimize suppression costs regardless of any other outcome, for example—the agencies could select a mix of firefighting assets and fuel reduction options predicted to maximize their ability to achieve that measure and consider the other measures only to help them choose between different mixes with similar outcomes for the most important measure. The agencies could also group two or more measures as more important than the others, or they could identify desired target levels for each measure and select the mix of firefighting assets projected to come closest to these targets. Senior officials will be able to use FPA to explore the modeled effects of weighting the measures differently—in effect, to evaluate the trade-offs associated with weighting any particular measure more heavily than the others—as well as to identify alternatives with high performance scores regardless of the weights ultimately selected.

The agencies will also need to determine whether the relative importance of the five measures is the same across different geographic regions of the country. Some officials and scientists involved with developing FPA have questioned whether applying a single weighting system across the country would accurately reflect national priorities or whether it is appropriate to emphasize different measures in different locations. For example, protecting the wildland-urban interface might be the most important consideration in some parts of the country, but reducing the likelihood of intense fires or protecting endangered species habitat might be more important elsewhere. FPA officials said that the model could perform this type of analysis, and agency budget officials said they would consider different approaches to weighting the measures once FPA was completed and the field units had submitted their different combinations of firefighting assets for analysis.

The leaders of FPA's science team told us that this new approach addressed specific concerns they had with the old approach. First, they said that using multiple measures to evaluate different mixes and locations of firefighting assets—rather than a single measure as in the old approach—better reflected the complexity of wildland fire management. Second, they said that the new approach is to analyze many more potential fire scenarios, thus evaluating the asset alternatives' predicted performance across a broader range of conditions than in the old approach.[10] Third, they said that because the old approach relied on weighting the relative importance of acres, they were concerned that different units would weight similar resources differently, thus preventing meaningful comparisons among units, or that some units might intentionally inflate the weights in an effort to gain advantage. Fourth, they said that because the new approach is to rely on alternatives developed by officials in the field, it can identify possible mixes and locations of assets that are likely to be more easily implemented than those identified through the old approach, which considered all possible combinations of assets typically available to local units and could suggest changes that might be unrealistic. Finally, they said that because the new approach allows field officials to identify the firefighting assets they would typically dispatch to fires burning in specific areas under certain conditions, it more closely follows how the agencies actually respond to fires.

The changes to FPA's modeling approach, however, were not among the recommendations stemming from the science review, which recommended that the agencies further test the initial model and improve it. But such testing and improvement of the initial model did not take place. The leaders of the science team told us that refining the initial model would not be useful, because the team had determined that the model was fundamentally flawed and a new approach was needed. Instead, the science team, in summer and fall 2006, developed five options for continuing to develop FPA and presented these options to the Wildland Fire Leadership Council, which selected one option in December 2006. This process was generally consistent with the management review, which recommended that an interagency science team examine the modeling approach FPA initially used.

Still, the agencies' rationale for making the changes to FPA's modeling approach was not fully documented, even though FPA is a major project whose outcome is expected to influence the allocation of billions of dollars. Although the science team's leaders told us they believed that the changes improved FPA, they provided no documents describing either the reasons for the changes or the process used to identify FPA's new approach. For example, a formal, documented comparison of the old and new approaches was never done; the science team's leaders told us they considered the relative strengths and weaknesses of the old approach and other possible approaches but did not document this consideration. In any event, each of the five development options the science team presented to the Wildland Fire Leadership Council included the same two fundamental changes in modeling approach. Without a formal, documented comparison of the old and new approaches, and without the opportunity to consider options that used other modeling approaches, the council lacked information that might have informed its choice.

FPA's Completion Has Been Delayed, but Costs Appear in Line with Previous Estimates

In addition, the changes apparently prevented the agencies from meeting their commitment to use preliminary FPA results beginning in 2006. Although FPA was not expected to be complete until late 2007, agency officials believed they would be able to make some use of its preliminary results in 2006. Accordingly, officials told congressional committee staff and OMB in early 2006 that the agencies would begin using FPA results that year to allocate their fiscal year 2007 funds and to develop their fiscal year 2008 budget requests.[11] Agency officials told us, however, that they subsequently decided not to use FPA's preliminary results because they did not believe it was prudent in light of the concerns that arose during the internal review. While it seems appropriate to delay using the model for budget decisions until concerns about its utility have been resolved, the agencies' position has been less than transparent; in August 2006—well after they realized that FPA would be undergoing substantial changes— they repeated their commitment to begin using FPA results in September of that year.

The agencies now expect that the FPA model will be completed in November 2008—about a year later than initially estimated—and that they will begin using FPA's results in 2009 to develop their 2011 budget requests, a delay of about 3 years from their initial goal of using preliminary results in 2006. When they began developing FPA in 2002, the agencies reported that FPA would be completed by the end of 2007. After the internal review, the agencies reported that a fully functional FPA system would be developed by June 30, 2008, and used in spring 2009 to inform the agencies' fiscal year 2011 budget requests. In spring 2008, the agencies repeated their commitment to this time frame. Agency officials attribute the delay in completing FPA to the project's complexity. When our review ended, agency officials said they expected fire planning units to begin using FPA in late 2008; about half the field units are expected to complete their alternatives by February 2009, with the remaining units completing their alternatives by June 2009. Meeting this time frame, however, will require the agencies to complete both the model and the guidance directing the field on how to develop the options the FPA model will analyze—both of which have experienced recent delays. Nonetheless, the agencies' Washington offices remained committed to using FPA results beginning in 2009. Agency field officials, however, have worried that the delay in completing FPA places an undue burden on the field by shortening the time available for planning units to develop their alternatives. Field officials also observed that senior agency officials have not clearly articulated how the results from FPA's first year would be used, although senior agency officials have stated that 2009 is to be a "learning year" and that they do not expect FPA to influence substantial changes to funding allocations in the first year.

The expected cost for completing FPA has been little affected by the substantial changes it has undergone since 2006. FPA's project development costs are expected to total about $43.9 million, according to an April 2008 estimate by the senior project manager responsible for FPA's budget.[12] This cost is generally in line with the agencies' previous estimates, particularly those developed after the agencies began to determine FPA's full scope (see table 1). Agency salaries and benefits, which were not included in yearly estimates of project development costs, represent an estimated $9.7 million in additional costs—for a total of about $53.6 million. According to the senior project manager, the agencies did not begin to develop FPA's second phase until 2005 and were still determining the scope of that phase when they submitted projected cost estimates in 2003 and 2004. The increase from 2003 to 2005 in the estimated cost for the second phase therefore reflects the agencies' better understanding FPA's scope and not a cost overrun, the project manager said. [13]

FPA Shows Promise in Achieving Some Objectives but Falls Short of Others, Although the Agencies Are Considering Changes That May Improve Its Performance

Although FPA is not yet complete and our review was limited, FPA shows promise in achieving some key objectives, including establishing a common, interagency budget framework that includes important wildland fire program activities. Nevertheless, FPA is unlikely to achieve all its key objectives, including the critical objectives of analyzing the effect over time of different funding allocation strategies and identifying the most cost-effective mix of firefighting assets. The agencies recognize that FPA will not fully meet all its key objectives in 2008 and are considering several changes that may improve its ability to meet certain objectives in the future. But because the modeling approach the agencies selected for FPA involves considerable discretion on the part of agency decision makers, transparency is particularly vital.

Table 1. Agencies' Cost Estimates for Developing FPA

Dollars in millions			
Year of cost estimate	**Phase 1 (fiscal years 2002-2006)**	**Phase 2 (fiscal years 2005-2010)**	**Total (fiscal years 2002-2010)**
2003	$11.9	$22.0	**$33.9**
2004	12.2	30.0	**42.2**
2005	12.1	36.2	**48.3**
2006	11.6[a]	31.2	**42.8**
Dollars in millions			
2007	11.6[a]	32.3	**43.9**
2008	11.6[a]	32.3	**43.9**

Source: GAO analysis of Forest Service data.

Note: Costs do not include salaries and benefits for all agency employees who worked on the FPA project. The senior project manager responsible for FPA's budget estimated these costs at $9.7 million.

[a]Actual, not estimated; the agencies completed phase 1 of FPA in 2005 at a cost of $11.6 million.

FPA Is to Provide the Foundation for an Interagency Framework for Analyzing Needed Firefighting Assets and Is to Examine Key Fire Management Program Activities and Objectives

If implemented as currently developed, FPA will provide the foundation for a single framework for the five federal agencies to develop their budget requests and allocate funds, a key objective. It is also likely to help the agencies achieve another key objective by analyzing the most important wildland fire management activities. The agencies have developed FPA so that it can recognize the presence of nonfederal firefighting assets that may be available to respond to fires on federal land—a third key objective—although the extent to which these assets will be included in the analysis is not yet clear. And finally, FPA should help the agencies move toward a fourth key objective—responding to wildland fires in ways that meet specific land and fire management objectives, rather than simply assuming that all fires should be suppressed as quickly as possible— although its ability to fully achieve this objective is likewise uncertain.

FPA Is to Provide the Foundation for an Interagency Budgeting Framework

As the agencies are developing it, FPA is to provide the foundation for a single framework for the five federal agencies to help develop their wildland fire budget requests and allocate their fire management funds, as envisioned in congressional guidance and the 2001 agency report—a significant step forward. In implementing FPA, officials are to work across agencies, both in the field and at headquarters. In the field, officials from each agency will need to work together to identify different mixes and locations of firefighting assets—information that will enable the FPA model to analyze the effect of different mixes of firefighting assets without regard to agency jurisdictional boundaries. At headquarters, agency officials are to work together to determine how to weight the five performance measures FPA incorporates to identify the best mix of firefighting assets.

FPA Is to Analyze the Three Most Important Fire Management Program Activities

FPA also substantially moves the agencies toward achieving another key objective by analyzing the three most important fire management activities: preparedness, fuel reduction, and suppression. FPA is to directly analyze preparedness and fuel reduction and then model the effects that varying investments in these activities might have on suppression costs.[14] Together, these activities constitute most of the agencies' overall fire management budgets.

To analyze the agencies' preparedness for wildland fires, FPA is to model the potential effect of wildland fire on communities and resources, depending on the mix and location of firefighting assets that would be stationed in an area. FPA is to consider historical fire occurrence and weather patterns to model the likelihood that a fire might occur in specific areas. Using an interagency database known as LANDFIRE to identify the fuel types and topography in the location where a fire is predicted to ignite,[15] FPA is to then model a fire's likely intensity and rate of spread. Finally, FPA is to identify the location of specific firefighting assets available for initial attack and, considering the fire's intensity and rate of spread, determine whether firefighters are likely to contain the fire before it grows too large and whether the fire is likely to damage communities or resources.

To analyze the effect of fuel reduction treatments within FPA, officials in the field are to begin by identifying the attributes of the fuel reduction treatments they most often undertake in their area, including vegetation type (such as trees, shrubs, or grasses) and the treatments' effect on vegetation density, height, and other characteristics. The FPA model is to then predict the effect of those treatments on fire behavior and compare the effectiveness of fuel treatments at reducing fire damage in different areas.

Finally, to analyze suppression costs, FPA is to consider different levels of investment in preparedness and fuel reduction and, for each investment level (including the mix and location of firefighting assets), estimate the number of fires likely to escape initial suppression efforts. For each such "large" fire, FPA is to use another model the agencies have developed to predict the cost of suppressing the fire on the basis of the costs from previous fires with similar characteristics, including fire size, fuel types, fire intensity, physical terrain, proximity to the nearest community, and total value of structures close to the fire. The costs of past fires with similar characteristics vary widely, however, which limits the model's ability to accurately predict suppression costs. Moreover, the model is based on historical costs, and since the agencies have recently begun emphasizing less aggressive strategies, it may not accurately predict suppression costs for fires.[16] The agencies are continuing to improve this model, however, which could improve the accuracy of the cost estimates.

FPA Is to Be Able to Analyze Nonfederal Firefighting Assets, but the Extent to which the Agencies Will Include These Assets in Their Analysis Is Unclear

Although the agencies are developing FPA to recognize the presence of nonfederal firefighting assets that may be available to respond to fires on federal land[17]—a key objective of FPA—the extent to which these assets will be included in the analysis is not yet clear. When officials in the field enter into FPA the different combinations of federal firefighting assets they would acquire for a given budget level, they can also include nonfederal firefighting assets that are stationed nearby, such as firefighters or fire engines belonging to state agencies or area communities. FPA is then to consider the availability of these nonfederal assets when it analyzes the effect of different combinations of federal assets on the five performance measures.

FPA officials recognize, however, that some nonfederal entities may object to federal agencies' including nonfederal assets in their analysis, for fear that doing so would lead to fewer federal firefighting assets stationed in certain locations, which in turn could lead to an additional workload for nonfederal entities in those locations. The inclusion of nonfederal assets raised significant concerns among nonfederal entities when the first FPA analysis was conducted in 2006. And while FPA guidance to planning units in the field generally directs them to include nonfederal assets, FPA officials acknowledged the likelihood that field units would receive "strong objections" to this direction from some nonfederal entities. Such objections might cause field units to omit nonfederal assets from the FPA analysis to satisfy the concerns of their nonfederal partners, with whom they must maintain relationships. FPA officials said they expect concerns from nonfederal officials to lessen over time, as those officials become more knowledgeable about how FPA operates. Ultimately, however, if agency planning units do not include nonfederal assets that may be available to respond to fires, FPA will model fewer firefighting assets than are actually present—and may therefore underestimate the effectiveness of a given set of federal assets. In addition, if some planning units include nonfederal assets and others do not, FPA's ability to identify the best combination of federal firefighting assets nationwide is likely to be compromised.

FPA Is to Consider Land and Fire Management Objectives, but Some Shortcomings Remain to Be Addressed

FPA should also help the agencies move toward achieving a fourth key objective—responding to wildland fires so as to meet specific land and fire management objectives, as suggested by the 1995 federal wildland fire management policy, rather than simply assuming that all fires should be suppressed as quickly as possible—although some agency officials have concerns about how well FPA will consider land management objectives. FPA should help the agencies move closer to this objective in two ways. First, officials in the field are to be responsible for identifying the number and type of firefighting assets they would typically dispatch to a fire that ignited in a particular location under particular conditions. The intent is to recognize that agency responses vary from fire to fire, and fire managers are more likely to dispatch more assets to a fire that threatens communities or highly valued resources or ignites under conditions conducive to rapid spread than to a fire ignited where it threatened few important resources or was unlikely to spread. The FPA model is to use this information to identify locations where stationing proportionately more firefighting assets might be helpful. Second, officials in the field are also to estimate the fire intensity beyond which resources in a particular area are likely to be damaged. In some areas, for example, officials might establish a relatively high intensity threshold to recognize that moderate, or even severe, fires might be acceptable, while in other areas—such as the wildland-urban interface—officials would likely determine that any fire is undesirable. In evaluating different mixes and locations of firefighting assets, FPA is to take into account this variation in acceptable fire intensity. In determining both the firefighting assets they would dispatch and the intensity threshold, field officials are expected to use information contained in local land and fire management plans, which the agencies are required to develop.[18]

Several issues, however, must be addressed for FPA to move the agencies more fully toward achieving their objective of responding to fires according to specific land and fire management objectives. First, one of the measures FPA is to use in evaluating alternative mixes and locations of firefighting assets is the predicted success of containing fires before they become large. Although containing fires when they are small is desirable in many circumstances, the agencies themselves have also recognized that their legacy of successful suppression has contributed substantially to the current increase in burned acres and fire intensity. As noted, it is not clear how the agencies will weight the relative importance of containing fires early (or indeed how they will weight any of the five measures) in FPA, but early guidance to the field indicates that early containment may be weighted heavily, which would keep FPA from fully recognizing the potential benefits of fire in some areas. Second, officials from the Fish and Wildlife Service and National Park Service have expressed concern that FPA is to evaluate the effects of reducing fuels solely by how the reduction affects the likelihood of a severe fire, without considering whether the fuel reduction treatment helps the agencies achieve broader land management objectives, such as improving the ecological condition of the land over time, as the 2001 report envisioned. A senior Fish and Wildlife Service official also noted that many wildlife refuges consist of small parcels of federal land interspersed among larger parcels of nonfederal land and that FPA is not designed to consider the effects of fragmented ownership.

Third, although FPA is to consider specific local land and fire management objectives that recognize that some areas are more important to protect than others, it will likely not allow the agencies to consistently identify the locations that are most important to protect from a national perspective. Within the five performance measures evaluating the effects of different mixes and locations of firefighting assets and fuel treatment options, FPA is to consider all acres as equally important, despite significant variation in the resources on those acres. For example, the agencies have established protection of the wildland-urban interface as one of their most important policy objectives, and FPA is to treat all interface acres identically, regardless of whether an acre contains one or several houses. Similarly, the agencies intend to increase the number of acres that are meeting fire and fuel management objectives, such as reducing the likelihood of uncharacteristically intense fires, and FPA is to consider all acres within this measure identically. For example, FPA is to consider an acre of a relatively common forest type, such as ponderosa pine, the same as a relatively rare type, such as giant sequoia—even though agency managers may place a much greater priority on the condition of a sequoia forest. As a result, FPA will not likely allow the agencies to give high priority to meeting objectives in particularly important or rare areas. FPA is also to predict the percentage of fires likely to be contained in initial attack. In evaluating the effect that different mixes and locations of firefighting assets have on this measure, however, FPA is to weight all fires equally, regardless of the fires' potential to damage communities or valuable natural or cultural resources. Agency officials analyzing FPA results may therefore consider it more important to try to contain multiple fires that do not pose a great threat than to try to contain a single fire that does. The presence of the other measures helps to mitigate this shortcoming, because if an uncontained fire damages communities or valuable resources, the agencies' ability to meet the other objectives will be compromised. The relative weights of the five measures, however, have not yet been determined, and it is not clear how the measures' interactions will play out.

One of the five measures the agencies will ultimately use to evaluate different mixes and locations of firefighting assets specifically considers resources the agencies regard as highly valued, which could improve the agencies' ability to identify some of the most important resources to protect. Nevertheless, FPA would still consider all acres within a particular performance measure identically and therefore not recognize that it is more important to protect some acres than others. In August 2008, the agencies decided to include only two types of resources in this measure in their 2009 analysis: municipal watersheds and habitat for some endangered species. Senior officials from the four Interior agencies, however, have criticized the approach the agencies are developing for FPA to consider highly valued resources because it does not sufficiently consider their agencies' land management objectives.

As Designed, FPA Will Not Achieve All Its Key Objectives, Including Examining the Effects over Time of Differing Funding Allocation Strategies and Identifying the Most Cost- Effective Mix of Firefighting Assets

Even though FPA is likely to achieve several of its key objectives, it is unlikely to help the agencies achieve others. In particular, the modeling approach the agencies are taking has limited ability to examine the effects over time of different funding allocation strategies and is unlikely to allow them to identify the most cost-effective mix of firefighting assets. Other aspects of FPA, including the lack of an external peer review of the model, may complicate its further development and implementation.

FPA's Ability to Examine the Temporal Effects of Differing Funding Allocation Strategies Appears Limited

FPA was envisioned as a way to help the agencies determine the extent to which, in the short term, increasing or decreasing funding for fuel reduction treatments and responding more or less aggressively to fires would affect the expected cost of responding to wildland fires over the long term. Although FPA is to analyze funding for both preparedness and fuel reduction, its ability to evaluate the trade-offs associated with increasing or decreasing one of these activities appears to be limited to short-term effects. Spending funds to reduce fuels, however, is generally considered a long-term investment, one whose value increases over time as more of the landscape is treated. If FPA considers only short-term effects, it may underestimate the benefit of reducing fuels and may lead the agencies to place greater emphasis on suppressing fires than warranted—with potentially far-reaching consequences.

FPA officials told us in September 2008 that they were working with the interagency science team to develop an approach that would allow FPA to better analyze the long-term effect of reducing fuels; the officials expected to incorporate this approach into FPA by November 2008.[19] Because the agencies had begun to develop this approach only toward the end of our review, we were unable to evaluate it. On the basis of our discussions with FPA officials and members of the interagency science team, and from our review of the limited documentation describing the approach, it appears that FPA's ability to help the agencies achieve this objective will be limited.

Moreover, FPA is unlikely to examine the effects over time of different firefighting strategies. Since adopting the 1995 fire management policy, the agencies have increasingly emphasized appropriate management response. FPA is to recognize that fire managers choose to respond less aggressively in some cases; for example, it is to allow field officials to model dispatching fewer firefighting assets to fires that are unlikely to threaten important resources. Fires responded to less aggressively are likely to burn many more acres than fires suppressed quickly. Less aggressive strategies may therefore reduce fuels on more acres—which in some cases could lower the risk from future large fires. FPA does not recognize this benefit, however, and will therefore be unable to help agency officials understand how responding less aggressively now may reduce the size and intensity of fires later, which could in turn help the agencies protect communities and resources and lower the cost of suppressing fires.

More broadly, FPA's limited ability to examine the effect over time of reducing fuels and implementing appropriate management response could also limit its ability to help the agencies develop a long-term, cohesive strategy for responding to wildland fires. We have long recommended that the agencies develop a cohesive strategy identifying available long-term options and associated funding for reducing hazardous fuels and responding to wildland fires.[20] Such a strategy is fundamental if the agencies and Congress are to fully understand the potential choices, and associated costs, for addressing wildland fire problems. The agencies have consistently concurred with our recommendation,[21] and agency officials cited FPA as a key step in enabling them to develop a cohesive strategy. In its current state of development, however, FPA lacks important capabilities to help inform strategic decisions about how to invest the agencies' limited funds.

FPA Is Unlikely to Allow the Agencies to Identify the Most Cost-Effective Mix of Firefighting Assets

A primary objective of FPA, established by the 2001 agency report, is to identify the most cost-effective fire management program for a given budget. Accomplishing this objective requires the agencies to define the fire management objectives they are trying to achieve and then to identify the combination of fuel reduction treatments and suppression strategies, including the best mix and location of firefighting assets, that would result in the most effective use of program funds. The modeling approach the agencies are using in FPA, however, does not allow the agencies to meet this objective. Rather than analyzing all possible combinations of assets typically available to local units, as well as fuel reduction and fire suppression strategies, to identify the most cost-effective combination, the approach the agencies are taking allows them to compare only a limited number of asset mixes and firefighting strategies (including fuel reduction options) to determine whether one mix of assets and strategies is more or less cost-effective than another. Because FPA is to compare only a limited number of alternatives, the evaluated alternatives are unlikely to include the most cost-effective mix of assets nationwide. Further, because the evaluated alternatives are likely to reflect minor variations in budget levels (e.g., plus and minus 5 percent or 10 percent of the prior year's budget for each planning unit), the present FPA approach is likely to generate results that differ only incrementally from the asset mixes and strategies already in place, rather than evaluate whether significantly different alternatives could yield significantly better results.

Agency officials, including key scientists involved in FPA development, told us they believed that although the modeling approach has not been designed to identify the single most cost-effective mix of firefighting assets, FPA would nevertheless provide useful information to help the agencies develop their budget requests. In fact, several officials told us they preferred the flexibility currently built into FPA, which allows them to consider multiple potential budget scenarios—what one official termed a "family of solutions"—over the rigidity built into the old approach, which resulted in a single solution. Officials told us that it would be unrealistic to expect that the complexities of wildland fire management could be modeled accurately enough to yield a single solution that is truly optimal and that by examining multiple possible budget scenarios developed by officials in the field, FPA's new approach would yield results that would be "among the most cost-effective solutions," according to one official. Nevertheless, it is not clear that examining only a small number of alternatives for each planning unit will generate results that are among the most cost-effective, particularly given current guidance to the field to consider only slight variations from current funding levels when developing alternatives.

Moreover, in analyzing trade-offs among different mixes and locations of firefighting assets, FPA is to consider only those assets that are stationed at individual management units, not those that are centrally located and under regional or national control. These central assets, which include large air tankers and helicopters and many of the most qualified firefighters, are some of the agencies' most costly, representing about $200 million of their budgets, according to agency estimates. The agencies use these assets in two ways: to assist local units with initial attack on small fires and to help suppress large fires. FPA is to consider the presence of these central assets when analyzing the likelihood that firefighters will be able to contain a fire during initial attack—important because otherwise the model would suggest that more firefighting assets would be needed at local planning units. FPA is to consider the number of centrally located assets as a given, however—that is, as a fixed input to the model, not a variable—rather than analyze the effects of changing the number of centrally located assets or the proportion of assets under local or national control. As a result, the model is not likely to determine the effect of changing the number or type of these assets on the agencies' firefighting abilities and costs, thus further limiting its ability to identify the most cost-effective mix of assets nationwide.

Other Aspects of FPA Complicate Further Development and Implementation

Although FPA is a new system, it will rely on many data sources, models, and systems the agencies developed earlier, some of which have known shortcomings. For example, FPA is to use data from the LANDFIRE system to identify the fuel types across the country; yet the accuracy of LANDFIRE data has been questioned, as has the frequency with which the system will be updated to recognize changes in fuel conditions over time due to insect outbreaks, large wildland fires, or other disturbances. Over the past several years, FPA and LANDFIRE project officials have worked together to develop a process to update LANDFIRE data, which should benefit FPA. It is too early to tell how effective the planned LANDFIRE improvements will be.

To predict how quickly a fire may spread in different fuel, weather, and geographical conditions, FPA is to use the results of FSPro, a fire growth model developed by Forest Service scientists. Fire officials have recognized that the spread rate predicted by the model is not always consistent with the rate of spread they observe during real fires. To help compensate for this difference, the FPA model is to allow field officials to calibrate the data used by FSPro to model spread rates so that they more closely reflect conditions typically observed in a particular area. It is not clear, however, how the agencies will ensure that the calibrations are made consistently or what the effects may be on the mix of firefighting assets FPA identifies as most appropriate.

Even with their known shortcomings, some of FPA's component elements are well-established applications that have been used by wildland fire managers for many years and, in some cases, are based on peer-reviewed science. The FPA model as a whole, however, including its component parts, has not been externally peer-reviewed. This lack comes in part because documentation on FPA's development and capabilities has not been sufficiently developed to allow for peer review; instead, according to agency officials, project staff have been devoting time to model development. Until the model is peer-reviewed—including validation that the overall logic is sound, the methods used are state of the art, the results are consistent with empirical evidence, and the system is adequate for its intended purpose—neither the agencies nor outside parties will have a full understanding of FPA's strengths and limitations or know how much confidence they should place in the model's analysis. A peer review, moreover, may identify limitations not revealed by our review.

Finally, the utility of FPA in identifying the best mix of firefighting assets will depend heavily on the alternatives and data developed by officials in the field, but some field officials expressed concerns about this component of FPA. For example, some field officials are concerned that they will receive little training on how to use FPA, which may prevent them from developing the most realistic options, and that the time needed to enter data into FPA and develop alternatives for national consideration will substantially increase their workload. Other officials told us that they are concerned that field staff may try to "game" FPA in an attempt to get the model to identify their area as needing more assets. Staff could, for example, develop a less-effective alternative for their low-budget scenario to make their mid- or high-budget scenarios appear more effective; similarly, staff may find it expedient to develop alternatives that ensure that each agency in their planning unit gains or loses comparable quantities of firefighting assets in order to promote equity among the agencies, rather than develop alternatives likely to best protect important resources but which might affect one agency more than another. Senior agency officials told us that gaming is a concern with any budgeting system and that they are planning to establish a two-stage process to review field submissions. In the first stage, officials from other field units would review the alternatives to ensure that interagency guidance was followed and information entered correctly; in the second, regional officials would review the results to ensure they met regional priorities. The exact steps this review process would follow, however, have not been determined, so it is not yet clear whether this process will ensure that only appropriately developed alternatives are submitted.

The Agencies Are Considering Changes That May Increase FPA's Ability to Meet Certain Key Objectives, Although These Changes Do Not Address All Shortcomings

Senior agency officials told us they recognize that in 2008 FPA will not fully meet all its key objectives but said they are considering making several changes that may improve its ability to meet certain key objectives, including the following:

- submitting FPA for an external peer review;
- continuing to develop FPA's process for identifying the resources the agencies consider to be highly valued and
- working with the interagency science team to improve how FPA is to consider land management objectives, such as improving the ecological condition of the land over time, when evaluating the benefits of reducing fuels.

Although these steps have the potential to move FPA closer to meeting some of its key objectives, it is too early to determine how successful they will be. Moreover, these steps do not address all shortcomings we or others have identified, and taking these or other steps to improve FPA will carry an additional cost, which is not included in current agency estimates. The approximately $54 million estimated cost for FPA includes basic operation and maintenance through fiscal year 2010 but, according to agency officials, does not include funds to make the above improvements.

The Approach Selected for FPA Increases the Importance of Transparency in Decision Making

The approach the agencies have taken in developing FPA allows for considerable discretion on the part of agency decision makers in three key areas: determining the relative importance (that is, the weights) of the five performance measures used to evaluate locally developed alternatives; using FPA results in combination with other information to develop agency budget requests; and using FPA results, likewise in combination with other information, to allocate funds to the field. Although it is important that decision makers have the flexibility to consider various options, that same flexibility makes it essential for the agencies to ensure that these processes are fully transparent. Otherwise, Congress, the public, and agency officials cannot be assured of fully understanding the rationale behind decisions or FPA's role in them. Although any changes to the existing allocation of funds among agencies or across different geographical areas are likely to be incremental at first, the agencies could consider larger funding reallocations as their understanding of FPA increases—which would make transparent decision making even more important.

First, as noted, the extent to which any particular alternative or set of alternatives is considered cost-effective will depend on the relative importance assigned to the five performance measures, including any variation in their relative importance in different regions of the country. FPA officials and the leaders of the interagency science team said that FPA is being designed to allow for the agencies to evaluate different weighting schemes, which senior agency officials referred to as "exploring the decision space." Others, however, have raised concerns that the flexibility inherent in setting weights for the different performance measures will allow the agencies to manipulate these weights until they reach a predetermined outcome. Without understanding the weights assigned to each measure and the rationale for assigning those weights— that is, without transparency in this process—Congress and others will find it impossible to understand and evaluate the reasonableness of FPA's results, and skepticism about FPA's usefulness will be difficult to quell.

Second, senior agency officials emphasized that, despite its importance, FPA will not be the sole determining factor in developing their budget requests and allocating appropriated funds; rather, senior agency officials would consider FPA results along with other information and exercise managerial discretion in making these decisions. Agency officials said, for example, that they would continue to involve national and regional officials from the various agencies to help ensure that their budget requests reflected differences in priorities among the agencies or regions, although they recognized that this process might lead FPA results to be used differently by different agencies or in different regions. Although considering these factors is important, as with the setting of the weights, it will also be important for the agencies to clarify the additional factors beyond FPA that they consider in developing their budget requests, so that Congress and others can understand FPA's role in the process.

And third, once Congress has appropriated funds to the agencies, it is not clear how the agencies will use FPA to help allocate these funds to the field. If one agency allocated funds differently than suggested by FPA—or if one agency's field unit acquired a different mix of assets than it modeled—it could affect the other agencies' ability to protect important resources, as well as the overall effectiveness of the agencies' fire management program. Agency officials said they intended for each agency to consider FPA results in allocating its funds and for field units to consider FPA results in acquiring firefighting assets. They also said they would not decide how much to deviate from the allocation suggested by FPA until they had begun to analyze the first year's results. Officials also said that it is important to recognize that more than 2 years could elapse between field units' developing their alternatives and Congress's appropriating funds on the basis of that information—and that priorities could change substantially in the interim, leading the agencies to allocate funds differently than suggested by FPA. Agency officials also said that, in addition to FPA results, they would consider specific congressional earmarks and appropriations guidance when allocating funds. Moreover, the agencies have existing systems outside of FPA for allocating fuel reduction funds, which they have been working in recent years to improve.[22] As of November 2008, agency officials did not know how they would consider the information from FPA in relation to the agencies' other systems in allocating fuel reduction funds.

CONCLUSIONS

As fires become more severe and development in fire-prone lands continues, the Forest Service and Interior agencies face difficult decisions about how to best protect the nation's communities and natural and cultural resources. In particular, the agencies must determine the best mix and location of firefighting assets to respond to wildland fires, and they must balance the need to spend money preparing for and fighting fires against the need to invest in reducing potentially hazardous fuels so as to lower both the cost of suppressing future fires and the risk to communities and important resources. Complicating these decisions, our nation's longterm fiscal challenges have constrained agency budgets, simultaneously limiting available choices and making it even more important to spend funds efficiently and effectively. The agencies believe that FPA can be a useful tool in making these difficult choices, which will drive billions of dollars in federal expenditures each year and directly affect millions of citizens living in fire-prone areas. By establishing an interagency budget framework that analyzes trade-offs among the most important fire management program activities, FPA represents an important first step in improving the agencies' cost-effectiveness.

Achieving the full potential of FPA, however, will depend on the extent to which the agencies improve FPA's ability to live up to the promises that were made on its behalf—namely, that it would allow the agencies to develop rational budgets and allocate funds in a way that maximizes the agencies' ability to manage wildland fire. Living up to these promises presents a daunting challenge, given the inherent difficulty of modeling the complexities and uncertainties of wildland fire and given that FPA remains a work in progress. Nevertheless, an early assessment of the model's capabilities raises several issues. The overall modeling approach the agencies have chosen does not allow them to identify the most cost- effective mix and location of firefighting assets, one of FPA's key objectives. Moreover, without improvements, FPA will be unable to identify, from a national perspective, the most important resources to protect or the relative priority of different values at risk; to evaluate the effect of different investments in fuel reduction treatments and firefighting strategies over time; or to analyze the effect of changes in the number of aircraft and experienced firefighters that are under regional or national control. Without such improvements, the agencies will continue to lack important information on which to base decisions about how best to allocate scarce funds. Further, the agencies have not yet determined how they will weigh the relative importance of FPA's five performance measures or exactly how they will use FPA to develop their budget requests and allocate funds. Given the importance of and the uncertainty surrounding these decisions, Congress—as well as the agencies and other interested parties—would benefit if these fundamental budget decisions were made in a transparent manner. And finally, an external peer review by an independent entity, such as the National Academy of Sciences, would achieve one of FPA's objectives and help the agencies identify the strengths and limitations of the model, which could increase confidence in their decisions and help them make needed changes more quickly.

RECOMMENDATIONS FOR EXECUTIVE ACTION

We recommend that the Secretaries of Agriculture and the Interior take four actions to improve their agencies' abilities to develop their budget requests and allocate funds using FPA.

First, to improve the FPA model's ability to identify needed firefighting assets and the best locations for these assets—and recognizing that developing FPA will be an iterative process that will require the agencies' continued effort to improve—we recommend that the Secretaries of Agriculture and the Interior direct the agencies to develop a strategic plan for the continued development of FPA, which would (1) include an evaluation of FPA's ability to meet its key original objectives; (2) identify ways to improve the model to better meet these objectives; (3) clearly state whether the agencies believe any of the original objectives are no longer appropriate, and why; and (4) identify the steps the agencies plan to take to improve FPA and the expected time frames and associated budget needs for completing these steps. To allow the agencies sufficient time to identify issues that may arise as they implement FPA, the Secretaries of Agriculture and the Interior should submit this plan to Congress no later than September 30, 2010. In particular, we believe that the strategic plan should, at a minimum, address ways to improve FPA's ability to

- evaluate different mixes and locations of firefighting assets, so that FPA recognizes the relative priority of different values at risk when assessing how best to protect the wildland-urban interface and increase the number of acres meeting fire management objectives;
- identify the most highly valued resources, such as endangered species habitat or important cultural sites, that the agencies seek to protect;
- model the effects over time of different investments in fuel reduction treatments and firefighting strategies on the cost of suppressing future wildland fires; and
- analyze trade-offs between increases and decreases in firefighting assets that are under national or regional control.

Second, we recommend that the Secretaries of Agriculture and the Interior report annually to Congress on (1) their progress in completing the steps outlined in the strategic plan for the continued development of FPA and (2) FPA's ability to meet each of its key objectives.

Third, to increase agency transparency in using FPA to develop their budget requests and allocate funds, we recommend that the Secretaries of Agriculture and the Interior report annually to Congress on FPA's role in the budget development and allocation process. This chapter should include, at a minimum, information on (1) how the agencies weighted the measures FPA uses to evaluate different mixes and locations of firefighting assets and the rationale for those weights, (2) how FPA results were used in conjunction with other information in developing the agencies' budget requests, and (3) the extent to which the agencies' funding allocations to their field units reflected the FPA results for a given year.

Fourth, to increase Congress's and the agencies' understanding of the strengths and limitations of FPA—including the extent to which it achieves the key objectives envisioned by the 2001 report—and to fulfill one of the original objectives established for FPA, we recommend that the Secretaries of Agriculture and the Interior direct the agencies to submit the FPA model to external peer review. This review should be initiated as soon as FPA is complete enough to allow for a thorough review, but no later than November 2009, so that its results can inform decisions about how FPA may be improved and the extent to which additional funding should be provided to the project.

AGENCY COMMENTS AND OUR EVALUATION

In written comments on a draft of this chapter, the Forest Service and Interior disagreed with our finding that FPA is unlikely to allow the agencies to identify the most cost-effective mix of firefighting assets. The Forest Service commented that it fundamentally agreed with our recommendations and described the steps the agency intended to take in addressing them, but we do not believe that the steps outlined in the letter are specific and transparent enough to meet the intent of our recommendations. Interior disagreed with our recommendation that the agencies develop a strategic plan for the continued development of FPA but concurred with our other recommendations.

The Forest Service and Interior commented that they believe FPA will allow them to meet the goal of cost-effectiveness. As their letters state, we previously discussed our conclusions on this issue with the agencies but did not resolve the differing points of view. As stated in our report, FPA compares only a limited number of mixes of firefighting assets and firefighting strategies, and the alternatives it evaluates are likely to reflect only minor variations in budget levels. Given this structure, we continue to believe that FPA is unlikely to allow the agencies to identify the most cost- effective location and mix of assets and strategies nationwide—an objective the agencies themselves established in their 2001 report. In their responses, both agencies raised questions about this objective. The Forest Service's comments seek to invalidate the objective altogether, stating that identifying the single most cost-effective mix of assets and strategies is not realistic. Interior did not question the validity of the objective but stated that the approach FPA is taking is more realistic than the approach the agencies had taken when they first began developing FPA. While we are not altering our conclusion that FPA's current approach will likely keep the agencies from identifying the most cost-effective solution, we are modifying our first recommendation to state that in the strategic plan, the agencies not only identify ways to improve the model to better meet FPA's original objectives, but also clearly state whether they believe any of the original objectives are no longer appropriate—and, if not, why not—in order to ensure that Congress and other interested parties are fully informed about what they can reasonably expect from FPA.

Regarding our recommendation that the agencies develop a strategic plan for the continued development of FPA, the Forest Service concurred with our recommendation and stated that it has a strategy for completing FPA, although it is not clear from the letter whether this strategy is or will be articulated in a written document directly addressing the elements of our recommendation. In contrast, Interior disagreed with this recommendation, stating that developing a strategic plan would delay the deployment and increase the cost of FPA. Regarding Interior's position, we are not suggesting that the agencies delay implementing FPA until they have developed the strategic plan we recommend; rather, we believe that such a plan can be developed concurrently with implementation and in fact may benefit from incorporating lessons learned during early use of FPA. More broadly, because of FPA's importance and the concerns about its development—including the questions raised in our review about its ability to meet its key objectives—we believe it is important for the agencies to create a strategic plan for FPA's continued development that directly and transparently evaluates FPA's ability to meet its original objectives, identifies ways to improve FPA to better meet those objectives, and identifies the steps the agencies plan to take to improve FPA. Given the agencies' comments about FPA's cost-effectiveness objective, however, we modified the language of our recommendation on developing a strategic plan, as discussed above.

The Forest Service and Interior generally agreed with our recommendations to report annually to Congress on the continued development of FPA and on FPA's role in the budget development process, and to submit the FPA model to external peer review. The Forest Service, however, also provided clarifications on two of our recommendations that did not appear to be fully responsive in terms of the amount of information and transparency we believe is warranted. Specifically, in response to our recommendations that the agencies report annually to Congress on (1) their progress in completing the strategic plan for FPA's continued development, and on FPA's current ability to meet each of its key objectives, and (2) FPA's role in the agencies' budget development and allocation process, the Forest Service stated that it has always been—and will continue to be—responsive to congressional requests for information and that it would include information on FPA's role in budget development and allocation in its annual budget requests. We are not convinced, however, that this approach will furnish Congress with the consistent, transparent, and complete information we believe it needs—particularly given FPA's importance in helping the agencies manage their $3 billion wildland fire program and the concerns about its development. We continue to recommend, therefore, that the Secretaries of Agriculture and the Interior prepare an annual report to Congress about the status of FPA's development and how the agencies have used FPA to help develop their budget requests and allocate funds. The Forest Service's and Interior's letters are reprinted in appendixes II and III, respectively, along with our evaluation of specific comments.

Sincerely yours,

Robin M. Nazzaro

Robin M. Nazzaro
Director, Natural Resources and Environment

APPENDIX I: SCOPE AND METHODOLOGY

To determine how the agencies have developed the fire program analysis (FPA) budget-planning system to date, we reviewed agency documents from each stage of FPA's development. To identify the key objectives originally established for FPA, we reviewed congressional committee and Office of Management and Budget guidance to the agencies; a 2001 report, commissioned and later adopted by the agencies, that established the vision, key objectives, time frames, and rationale for what FPA was intended to accomplish; the interagency memorandum of agreement and project charter that established FPA as an interagency project; and other agency documents. To further our understanding of the broader context for the shortcomings FPA was intended to address, we reviewed key agency documents, including the 1995 and 2001 federal wildland fire management policies, the national fire plan, and related documents. To identify changes the agencies made to FPA in 2006, and the reasons for those changes, we reviewed the report the agencies issued after their review of FPA's policy and scientific approaches; a response to that report prepared by those who had helped to develop FPA; and internal agency briefing materials about the changes. To identify the likely capabilities of FPA as the agencies have been developing it since 2006, we reviewed the draft interagency science team report that formed the basis for FPA's new modeling approach and numerous technical papers and other documentation describing particular aspects of FPA. To further our understanding of FPA's development at each of these stages, we interviewed Forest Service and Department of the Interior officials in Washington, D.C.; FPA project staff in Boise, Idaho; and agency officials in the field familiar with FPA. We also interviewed agency and other scientists who have helped develop FPA.

To determine the extent to which FPA meets its original objectives, we compared—to the extent possible—the capabilities of FPA as the agencies developed it with those envisioned in congressional committee guidance and the 2001 report. At the time of our review, however, substantial portions of the model remained incomplete, and the agencies had not sufficiently documented the model to allow a comprehensive evaluation. We therefore limited our review to a broad examination of FPA's various components and how they interact. We also interviewed senior agency officials, FPA project staff, agency field officials, and agency and other scientists to obtain their views on the extent to which FPA appears capable of meeting its key objectives, as well as possible changes that could improve the model's ability to meet those objectives.

We conducted this performance audit between September 2007 and November 2008 in accordance with generally accepted government auditing standards. Those standards require that we plan and perform the audit to obtain sufficient, appropriate evidence to provide a reasonable basis for our findings and conclusions based on our audit objectives. We believe that the evidence obtained provides a reasonable basis for our findings and conclusions based on our audit objectives.

United States Department of Agriculture | Forest Service | Washington Office | 1400 Independence Avenue, SW Washington, DC 20250

File Code: 1420/5100
Date:

Ms. Robin M. Nazzaro
Director, Natural Resources and Environment
United States Government Accountability Office
4441 G Street, N.W.
Washington, DC 20548

Dear Ms. Nazzaro:

Thank you for the opportunity to comment on the draft Government Accountability Office (GAO) report, GAO-09-68, "Wildland Fire Management: Interagency Budget Tool Needs Further Development to Fully Meet Key Objectives." The Fire Program Analysis (FPA) project is a very significant and challenging undertaking by the Federal wildland fire agencies. We were pleased the audit team discussed the complexities inherent in developing an interagency planning and budget system. While we generally view the audit as supportive, we respectfully disagree with GAO's conclusion that our approach hampers FPA from meeting the key objective of cost effectiveness. We discussed this with the audit team; they were receptive to the discussion and recommended that we document our concern.

As indicated, we worked closely with GAO, commenting and clarifying statements; however, not all of these have been reflected in key aspects of the draft report. We believe GAO has not accurately portrayed the system's ability to meet the cost effectiveness objective and actions we have taken to assure that FPA is a useful planning and budgeting tool. GAO takes exception to FPA system design modifications in 2006 that it says compromises the agencies' ability to fully achieve key goals. We strongly disagree that these modifications compromise cost effectiveness.

> GAO contends that the analytical approach developed by the agencies cannot identify the most cost effective mix and location of federal firefighting assets for a given budget, but rather can only compare the relative cost effectiveness of a small set of alternatives. This assertion fails to recognize or acknowledge several key points. First, the notion of a single most cost-effective solution is conceptual; it is not based in reality. Any reasonable alternative can be shown to be optimal under a hypothetical set of assumptions about fire occurrence, weather, and values placed on the consequences of individual fires. That same alternative would be judged inferior to other alternatives under differing sets of assumptions, even if the differences are minor. Furthermore, building an optimization model requires gross simplification of the system being modeled in order to calculate a solution. This simplification can lead to distorted or inaccurate representations of the relationships between actions and outcomes, which reduces the confidence placed in a given solution. These and other shortcomings of the classical optimization approach were manifest in the first phase of FPA and led the agencies to change their approach.

Caring for the Land and Serving People

Printed on Recycled Paper 

Note: GAO comments supplementing those in the report text appear at the end of this appendix.

See comment 1.
See comment 2.
See comment 2.

Ms. Robin Nazzaro 2

The analytical approach recommended by the Interagency Science Team and adopted by the agencies remains faithful to the goal of improving firefighting effectiveness. It also improves the ability to: A) address the uncertainty inherent in wildland fire due to random variations in weather, fuels, and topography; B) more realistically model fire behavior and the strategic and tactical choices made by wildland fire managers and; C) build upon the corporate intelligence gained from decades of firefighting by the agencies and their partners.

This approach will allow us to systematically evaluate alternative investment strategies and identify options that best reduce fire losses, improve ecological conditions and increase cost efficiency. The system is designed to explicitly address uncertainty and risk in predicting future wildland fires. A combination of simulation models and goal programming will array alternatives using quantitative performance measures that display inherent risks and trade-offs at both local and national levels. This approach is a more robust basis for modeling real-world complexities than the linear optimization approach originally used in FPA, while maintaining the ability to compare the performance and effectiveness of alternative funding decisions.

We would also like to offer comments relative to the system's ability to evaluate fuel reduction investments over time. GAO states that we are working "to develop an approach that would allow FPA to better analyze the long-term effect of reducing fuels" but that due to the development time frame they were unable to evaluate it, reaching the opinion "it appears that FPA's ability to help the agencies achieve this objective will be limited." GAO concludes that "without improvements FPA will be unable to evaluate the effect of different investments in fuel reduction and firefighting strategies over time." We want to reaffirm our commitment to ensure that the system is useful and that it supports, both near-term and long-term, fire planning and budgeting. To that end, our development and science teams are aggressively analyzing temporal modeling approaches for fuel treatments which could be released later this year. In addition, the performance metric "proportion of land meeting or trending toward the attainment of fire and fuels management objectives" will recognize and consider acres managed under the appropriate management response. The system now being deployed is within the scope approved by the Agency. However, the Forest Service has recognized the potential for a more comprehensive analysis of vegetation and fuel treatments that will address the concerns expressed by GAO.

GAO Recommendations - The report identifies four recommendations to improve FPA and its use in the budget process. The Forest Service fundamentally agrees with GAO's recommendations, but believes there are better alternative approaches for implementing three of the recommendations than those proposed by GAO.

Recommendation 1: The Secretaries of Agriculture and the Interior direct the agencies to develop a strategic plan for the continued development of FPA, which would include: 1) an evaluation of the strengths and weaknesses of FPA; 2) identify ways to improve the model to better meet its intended objectives, and; 3) identify the steps the agencies plan to take to improve FPA and expected time frames and associated budget.

See comment 4.
Enclosure not reprinted.
See comment 5.

Ms. Robin Nazzaro 3

Response - The Forest Service has a strategy for completing development and implementation of the FPA system consistent with the project's charter, and its associated plans. The Forest Service has recognized the need to implement FPA in an adaptive manner in FY 2009 through a staged approach that allows for system adjustments as experience dictates. This approach is both helping to avoid workload issues for the system and development of personnel and facilitating our ability to address system issues as they arise. The enclosed graphic displays the schedule and methodology for implementation across all Fire Planning Units. As we use the system and review and analyze its outputs, its strengths and weaknesses will be identified and documented through existing business processes. In addition, the planned external peer review will provide insight into FPA's strengths and weaknesses.

Major development of FPA will be complete in FY 2009 when the system will transition from development to operation and maintenance. The system's Operation and Maintenance Plan will provide for some enhancements to the system's models. In addition, the Forest Service has recognized other modeling components, such as a focused analysis of national resources or an expanded analysis of fuels and vegetative treatments, which could be useful and potentially provide a more comprehensive range of alternatives. These will be considered as the system is deployed and insights into outputs, and their utility, become known. The FPA Executive Oversight Group will consider these and other enhancements and provide guidance relative to their future inclusion and development.

Key aspects of these actions and activities will be conveyed in accordance with the Forest Service's response to Recommendations 2 and 3.

Recommendation 2: The Secretaries of Agriculture and the Interior report annually to Congress on: 1) their progress in completing the steps outlined in the strategic plan for the continued development of FPA; and, 2) FPA's ability to meet the key objectives initially established for it.

Response - The Forest Service agrees with informing Congress about the progress in implementing FPA and how the results of FPA are being used in agency decision processes. The agency has always been responsive to Congressional requests for information through informal briefings, responses to written questions, and formal testimony. The Forest Service will continue to respond to any requests for information by members and committees. In addition, the results and use of FPA information will be clearly highlighted in the Agency's formal annual budget requests to Congress.

Recommendation 3: To increase agency transparency in using FPA to develop their budget requests and allocate funds, the Secretaries of Agriculture and the Interior report annually to Congress on FPA's role in budget development and allocation process.

Response – Please see response to recommendation 2.

Ms. Robin Nazzaro 4

Recommendation 4: The Secretaries of Agriculture and the Interior direct the agencies to submit the FPA model to external peer review.

> **Response** - The Forest Service agrees with the recommendation. An external peer review is planned as part of the FY 2009 development and implementation strategy.

Please contact Sandy T. Coleman, Forest Service Assistant Director for GAO/OIG Audit Liaison staff, at 703-605-4699, with any questions.

We look forward to working with GAO in the future.

Sincerely,

ABIGAIL R. KIMBELL
Chief

cc: Sandy T Coleman, Clarice Wesley, Tom Harbour, Bill Breedlove, Rick Prausa

Appendix II: Comments from the Department of Agriculture Foest Serce

The following are GAO's comments on the Department of Agriculture, Forest Service's letter dated November 5, 2008.

GAO Comments

1. As the Forest Service's comment letter indicates, we have had extensive discussions with the agency on FPA's ability to identify the most cost-effective mix of firefighting assets, without resolving our differing points of view. As we describe in our report, FPA compares only a limited number of mixes of firefighting assets and firefighting strategies, and further, the alternatives it evaluates are likely to reflect only minor variations in budget levels. Given this structure, we continue to believe that FPA is unlikely to allow the agencies to identify the most cost-effective location and mix of assets and strategies—an objective the agencies themselves established in their 2001 report. Rather than directly contradicting our conclusion, however, the Forest Service's letter seeks instead to invalidate this objective altogether. The Forest Service commented that identifying the most cost-effective mix of assets and strategies is not a realistic objective and that FPA's current combination of simulation models and goal programming is a preferable approach. We did not compare the agencies' current approach with their initial approach, nor have we concluded whether one approach is more suitable or realistic than the other. Rather, in accordance with the objectives of our review, we simply evaluated the extent to which FPA as it is currently being developed is likely to meet the objectives originally established for it. While we are not altering our conclusion that FPA's current approach will likely not result in identifying the most cost-effective solution, we are modifying our first recommendation to suggest that the agencies clarify which of FPA's original objectives they believe are no longer appropriate and why. See comment 4 below.
2. The Forest Service stated that FPA remains faithful to the goal of improving firefighting effectiveness and that FPA's approach will provide a more robust basis for systematically evaluating alternative investment strategies. As discussed above, however, and as we noted in our draft report, the objective originally established for FPA was to identify the most cost-effective mix and location of firefighting assets and strategies, not simply to improve firefighting effectiveness.
3. The Forest Service's letter reaffirmed the agency's commitment to ensuring that FPA is able to support both near-term and long-term planning considerations in evaluating fuel reduction investments. In reaffirming this commitment, the Forest Service stated that FPA, as it is being developed, fulfills the scope that has been approved by the agency. This approved scope, however, has evolved during FPA's development and is not fully consistent with the objectives initially established for FPA. Our conclusions about FPA are based on our comparison of its current capabilities with the objectives originally established for it. We did not determine whether the

agencies were developing FPA in a manner that fulfilled the scope approved by the agencies in subsequent documents.

4. The Forest Service stated that it has a strategy for completing FPA "consistent with the project's charter and its associated plans." It is not clear from the letter, however, whether this strategy is, or will be, articulated in a written document directly addressing the elements of our recommendation. Because of FPA's importance, and the concerns that have arisen during its development, we believe it is important for the agencies to develop a single document that addresses these issues transparently. Given the agencies' comments, however, we modified our recommendation to suggest that the agencies use this plan not only to identify ways to improve the model to better meet FPA's original objectives, but also to clearly state whether they believe any of the original objectives are no longer appropriate, and why, in order to ensure that Congress and other interested parties are fully informed about what they can reasonably expect from FPA.
5. The Forest Service stated it would respond to any requests for information by Congress and would highlight how FPA's results were used in the agency's annual budget request. We are not convinced, however, that this approach will provide Congress with the consistent, transparent, and complete information we believe it needs— particularly given FPA's importance in helping the agencies manage their $3 billion wildland fire program and the concerns about its development. We continue to recommend, therefore, that the Secretaries of Agriculture and the Interior prepare an annual report to Congress about the status of FPA's development and how the agencies have used FPA to help develop their budget requests and allocate funds.

APPENDIX III: COMMENTS FROM THE DEPARTMENT OF THE INTERIOR

THE ASSOCIATE DEPUTY SECRETARY OF THE INTERIOR
WASHINGTON

NOV 5 2008

Ms. Robin M. Nazzaro
Director, Natural Resources and Environment
Government Accountability Office
441 G Street, NW
Washington, D.C. 20548-0001

Dear Ms. Nazzaro:

We appreciate the opportunity to review and comment on the draft Government Accountability Office report entitled, "*Wildland Fire Management: Interagency Budget Tool Needs Further Development to Fully Meet Key Objectives*," (GAO-09-68). The Fire Program Analysis project is a very significant and challenging undertaking by the Federal wildland fire agencies. We were pleased the audit team engaged in numerous and constructive discussions to understand the complexities inherent in developing an interagency planning and budget system. While we view the audit as supportive of our effort, we respectfully disagree with GAO's conclusion that our approach hampers FPA from meeting key objectives. We discussed this concern with the audit team; they were receptive to the discussion and recommended that we document our concern.

We are concerned that our comments and clarifying information have not been reflected in the report. In particular, we believe that FPA will allow us to meet the cost effectiveness objective and that actions we have taken will ensure that FPA will be a useful planning and budgeting tool. The FPA will be useful in meeting the cost effective objective and be a useful budgeting and planning tool, because FPA is an interagency analysis that is based on collaboration at the local level. The results of the collaborative analysis process will display efficiencies and effectiveness identified at the local planning level. We also believe that the 2006 system modifications support the goal of cost effectiveness. The cost effectiveness is based on quantitative performance measures that display trade-offs locally and nationally between various budget levels. The FPA system allows the local planning unit to determine the most effective mix of resources and fuel treatments at various budget levels. The FPA does not provide one finite answer but instead evaluates the changes in modeled effectiveness for each investment level analyzed.

We believe that one of FPA's greatest strengths is the capability to evaluate alternative investment strategies and identify options that best reduce fire losses, improve ecology conditions, and increase cost efficiency. This ability to array alternatives that display risks and trade-offs at the local and national levels provide a robust basis for comparing

Note: GAO comments supplementing those in the report text appear at the end of this appendix.

potential performance and cost effectiveness will optimize the use of resources. We believe this is a much more realistic approach then the single-dimensional optimization approach for the most cost effective mix and location of firefighting asset for a given budget. Our Interagency Science Team has recommended the more analytical approach that we are now developing.

With respect to the system's ability to evaluate fuel reduction investments over time, it is important to note that our development and science teams are analyzing temporal modeling approaches for fuel treatments, which may be released later this year. We agree with GAO that there is potential for a more comprehensive analysis of vegetation and fuel treatments.

Our comments on the recommendations are as follows:

Recommendation 1—The Secretaries of Agriculture and the Interior develop a strategic plan for the continued development of FPA, which would (1) include an evaluation of the strengths and weaknesses of FPA, (2) identify ways to improve the model to better meet its intended objectives, and (3) identify the steps the agencies plan to take to improve FPA and expected time frames and associated budget needs for completing these steps.

Response— The results of the reviews of FPA that were conducted by the science team and the management team support the current course of action. The system is expected to provide results that can be used in 2009 and to be available in time to develop the fiscal year 2011 budget request. The fire community is anxiously awaiting access to this system that will give them new tools to use in decision-making and improve their ability to allocate resources in a cost effective manner. Development of a strategic plan will further delay system deployment and result in increased costs. We would prefer to continue forward with the planned, staged approach that allows for adaptive use and modification. We believe that this approach will allow us to ensure that the system optimizes capabilities for evaluation of different mixes and locations of firefighting assets, protection of valued resources, modeling investments, and analyzing trade-offs in the allocation of firefighting assets. In addition, our planned external peer review is expected to further assist in identifying improvements to the system.

Recommendation 2— The Secretaries of Agriculture and the Interior report annually to Congress on (1) their progress in completing the steps outlined in the strategic plan for the continued development of FPA, and (2) FPA's current ability to meet each of the key objectives initially established for it.

Response— The Department concurs with GAO's recommendation to report annually to Congress on progress and achievement of objectives.

Recommendation 3—Increase agency transparency in using FPA to develop their budget requests and allocate funds by reporting to Congress on FPA's role in budget development and allocation process.

Response— The Department concurs with GAO's recommendation and will ensure that the results and use of FPA information is clearly depicted in the budget request to Congress.

Recommendation 4—Submit the FPA model to external peer review.

Response— The Department concurs with GAO's recommendation and is planning an external peer review.

We have closely coordinated our response with the U.S. Forest Service and hold concurrent views. If you have any questions or concerns, please contact Barbara Loving at the Office of Wildland Fire Coordination at 202-606-3108.

We look forward to working with GAO in the future.

Sincerely,

James E. Cason

James E. Cason

Enclosure

The following are GAO's comments on the Department of the Interior's letter dated November 5, 2008.

GAO Comments

1. As Interior's comment letter indicates, we have had extensive discussions with agency officials on FPA's ability to identify the most cost-effective mix of firefighting assets without resolving our differing points of view. Interior commented

that FPA will allow the department to meet FPA's cost-effectiveness objective by evaluating alternative investment strategies and identifying options that best reduce fire losses, improve ecological conditions, and increase cost efficiencies and that the agencies' current approach is much more realistic than the approach taken initially. As we describe in our report, FPA compares only a limited number of mixes of firefighting assets and firefighting strategies, and further, the alternatives it evaluates are likely to reflect only minor variations in budget levels. Given this structure, we continue to believe that FPA is unlikely to allow the agencies to identify the most cost-effective location and mix of assets and strategies—an objective the agencies themselves established in their 2001 report. And as noted in our response to the Forest Service's comments, we did not compare the agencies' current approach with their initial approach, nor do we conclude whether one approach is more suitable or realistic than the other. Rather, in accordance with the objectives of our review, we simply evaluated the extent to which FPA as it is currently being developed is likely to meet the objectives originally established for it. While we are not altering our conclusion that FPA's current approach will likely keep the agencies from identifying the most cost-effective solution, we are modifying our first recommendation to suggest that the agencies clarify which of FPA's original objectives they believe are no longer appropriate and why. See comment 3 below.

2. Interior commented that the agencies are continuing to analyze how FPA evaluates fuel reduction investments over time and may begin using a new modeling approach later in 2008. It is not clear from Interior's letter whether it believes the new approach will allow the agencies to meet FPA's original objective of modeling the effects over time of differing strategies for responding to wildland fires and treating lands to reduce hazardous fuels. Our review of the limited documentation describing this approach suggests that this approach is unlikely to allow FPA to fully meet this key objective.
3. Interior stated that developing a strategic plan for the continued development of FPA, as we are recommending, would further delay deployment and increase the cost of FPA. We recognize it is important for the agencies to continue to develop FPA, and we are not suggesting that the agencies delay implementing FPA until they have developed the strategic plan we recommend. On the contrary, we believe that such a plan can be developed concurrently with implementation and that the agencies may benefit from incorporating lessons learned during FPA's early use into the plan. In any event, our review raised questions about FPA's ability to meet certain of its key objectives, even with the changes the agencies are considering making to FPA—and because of FPA's importance, and the concerns about its development, we believe it is important for the agencies to develop a single document that directly and transparently evaluates FPA's ability to meet its original objectives and identifies ways to improve FPA to better meet those objectives. Given the agencies' comments, however, we modified our recommendation to suggest that the agencies use this plan not only to identify ways to improve the model to better meet FPA's original objectives, but also to clearly state whether they believe any of the original objectives are no longer appropriate, and why, in order to ensure that Congress and other interested parties are fully informed about what they can reasonably expect from FPA.

End Notes

[1] Departments of Agriculture and the Interior, *Federal Wildland Fire Management Policy and Program Review* (Washington, D.C., December 1995). This policy was subsequently reaffirmed and updated in 2001: Departments of the Interior, Agriculture, Energy, Defense, and Commerce; Environmental Protection Agency; Federal Emergency Management Agency; and National Association of State Foresters, *Review and Update of the 1995 Federal Wildland Fire Management Policy* (Washington, D.C., January 2001).

[2] Forest Service and Department of the Interior, *Developing an Interagency, Landscape- scale Fire Planning Analysis and Budget Tool* (Washington, D.C., November 2001).

[3] Departments of Agriculture and the Interior, *Fire Program Analysis: Scientific Review Team Report* (Washington, D.C., January 2006); and *Management Review Team Report of the Fire Program Analysis (FPA) Preparedness Module* (Washington, D.C., March 2006).

[4] Together, preparedness, suppression, and fuel reduction make up approximately 80 percent of the agencies' wildland fire management budgets. Other federal wildland fire program components include financial assistance to state foresters for fire management activities, research and development, and rehabilitating burned federal lands.

[5] Federal and nonfederal agencies have established a framework to share the costs of responding to fires that threaten both federal and nonfederal resources. See GAO, *Wildland Fire Suppression: Lack of Clear Guidance Raises Concerns about Cost Sharing between Federal and Nonfederal Entities*, GAO-06-570 (Washington, D.C.: May 30, 2006).

[6] The number of planning units established by the agencies has fluctuated over the course of FPA development. In this report, we refer to the 139 planning units in existence at the time our review ended but recognize that the actual number at any particular time may differ.

[7] Peer review is a process by which scientific research or technical projects are subject to an independent assessment by scientists not involved with the project who have knowledge and expertise comparable to that of the scientists whose work they review.

[8] The Wildland Fire Leadership Council consists of senior Agriculture and Interior officials, including the Agriculture Undersecretary for Natural Resources and Environment; the Interior Assistant Secretary for Policy, Management, and Budget; and the heads of the five federal firefighting agencies. Other members include representatives of the Intertribal Timber Council, the National Association of State Foresters, and the Western Governors' Association, and a local fire department chief.

[9] The size threshold for fires to be considered contained while small is to vary according to criteria established by officials in each planning unit, considering the circumstances under which they typically consider a fire in their area "escaped" and then request additional firefighting assets to help suppress it. This measure also includes the number of fires the model predicts would be averted because of the agencies' efforts to prevent human-caused fires.

[10] The program staff who helped develop the old approach told us that they had recognized the small number of potential fire scenarios in that approach limited its capabilities and that they were considering how to improve it, but the agencies determined that a new modeling approach was needed before they could make improvements.

[11] The agencies have provided brief updates on the status of FPA to Congress in their annual budget justifications, and have provided periodic briefings to congressional committee and OMB staff.

[12] These figures include the cost of developing FPA and operating and maintaining it through fiscal year 2010.

[13] Similarly, the project manager said that the decrease in estimated cost for the second phase from 2005 to 2006 was also due to a better understanding of the project's scope.

[14] FPA is also to help the agencies model their investment in preventing fires. The agencies carry out activities, such as increased law enforcement patrols and public education programs, intended to reduce the number of human-caused wildland fires. FPA is to predict the number of fires that would have started if not for the agencies' prevention activities.

[15] LANDFIRE is a geospatial data and modeling system designed to assist the agencies in identifying the extent, severity, and location of wildland fire threats to the nation's communities and ecosystems. At the time of our review, LANDFIRE data were not available for the eastern United States or for Alaska and Hawaii. FPA officials said that until LANDFIRE data are available nationwide, they are using other available data to provide similar information. FPA officials expect that LANDFIRE data will be available nationwide by 2009.

[16] We have previously reported limitations of the model the agencies use to predict suppression costs. See GAO, *Wildland Fire Management: Lack of Clear Goals or a Strategy Hinders Federal Agencies' Efforts to Contain the Costs of Fighting Fires*, GAO-07-655 (Washington, D.C.: June 1, 2007).

[17] In some cases, federal firefighting assets are also available to respond to fires on nonfederal land.

[18] We have previously reported on the status of the agencies' development of these plans. See GAO, *Wildland Fire Management: Update on Federal Agency Efforts to Develop a Cohesive Strategy to Address Wildland Fire Threats*, GAO-06-671R (Washington, D.C.: May 1, 2006).

[19] The interagency science team in 2006 proposed an option for developing FPA that might have helped the agencies to better achieve this objective, but the Wildland Fire Leadership Council did not approve this option out of concern that the agencies would be unable to complete it within the time and budget available. Interagency science team members told us they could, if directed, continue to develop that option and incorporate it into FPA later.

[20] GAO, *Western National Forests: A Cohesive Strategy Is Needed to Address Catastrophic Wildfire Threats*, GAO/RCED-99-65 (Washington, D.C.: Apr. 2, 1999); *Wildland Fire Management: Important Progress Has Been Made, but Challenges Remain to Completing a Cohesive Strategy*, GAO-05-147 (Washington, D.C.: Jan. 14, 2005); and GAO-06-671R.

[21] In 2008, however, we reported that the agencies had begun retreating from their commitment to develop a cohesive strategy. See GAO, *Wildland Fire Management: Federal Agencies Lack Key Long- and Short-Term Management Strategies for Using Program Funds Effectively*, GAO-08-433T (Washington, D.C.: Feb. 12, 2008).

[22] For information on the agencies' approaches to allocating fuel reduction funds, see GAO, Wildland Fire Management: Better Information and a Systematic Process Could Improve Agencies' Approach to Allocating Fuel Reduction Funds and Selecting Projects, GAO-07-1168 (Washington, D.C.: Sept. 28, 2007).

CHAPTER SOURCES

Chapter 1 - This is an edited, reformatted and augmented edition of a Congressional Research Service publication, Report RS22747, dated January 30, 2008.

Chapter 2 - This is an edited, reformatted and augmented edition of an United States Department of Agriculture, Forest Service Pacific Southtwest Research Station, Research Paper, PSW-RP-257 dated June 2008.

Chapter 3 - This is an edited, reformatted and augmented edition of a Congressional Research Service publication, Report RL30755, dated January 29, 2009.

Chapter 4 - This is an edited, reformatted and augmented edition of a Congressional Research Service publication, Report RL34517, dated June 2, 2008.

Chapter 5 - This is an edited, reformatted and augmented edition of a Congressional Research Service publication, Report RS21800, updated June 12, 2008.

Chapter 6 - This is an edited, reformatted and augmented edition of a United States Government Accountability Office publication, Report GAO-09-68, dated November 2008.

INDEX

C

D

E

F

G

H

I

R

S

T

U

V

W

Y

Z